DATE DUE

International Dictionary
of Broadcasting and Film

International Dictionary of Broadcasting and Film. 2d ed. By Desi K. Bognar. 1997. 268p. Focal Press, paper, $26.95 (0-240-80212-8).

Focusing on the technical side of motion picture production, these resources from Focal Press define terms like *gaffer, reference white, shooting ratio,* and *Panavision*. Elkins' more specialized book is illustrated. Acronyms and abbreviations abound. Bognar's broad reach makes it essential for any basic film reference collection.

International Dictionary of Broadcasting and Film

Desi K. Bognár

Focal Press

Boston Oxford Melbourne Singapore Toronto Munich New Delhi Tokyo

Focal Press is an imprint of Butterworth–Heinemann.

$\mathcal{R}$ A member of the Reed Elsevier group

∞ Recognizing the importance of preserving what has been written, Butterworth–Heinemann prints its books on acid-free paper whenever possible.

Library of Congress Cataloging-in-Publication Data
Bognár, Desi K. (Desi Kégl)
    International dictionary of broadcasting and film / Desi K.
Bognár.
      p. cm.
    ISBN 0-240-80212-8 (pbk. : alk. paper)
    1. Broadcasting—Dictionaries.   2. Motion pictures—Dictionaries.
  3. Cinematography—Dictionaries.   4. Video recording—Dictionaries.
  I. Title.
PN1990.4.B64   1995
791.4'03—dc20
                                     95–16675
                                     CIP

British Library Cataloguing-in-Publication Data
A catalogue record for this book is available from the British Library.

The publisher offers discounts on bulk orders of this book.
For information, please write:

Manager of Special Sales
Butterworth–Heinemann
313 Washington Street
Newton, MA 02158-1626

10 9 8 7 6 5 4 3 2 1

Printed in the United States of America

To
my wife Katalin, born in Budapest,
my son Istvan, born in New York City, and
my daughter Janina, born in Ibadan, Nigeria

Contents

Preface	**ix**
Acknowledgments	**xi**
Abbreviations	**xiii**
The Language: Terms A–Z	**1**
Appendix A: English–Metric Conversion Tables	**251**
Magnetic Tape Width Standards	251
Magnetic Tape Speed Standards	251
Phonograph Speeds	251
Film Length Standards	251
Negative, Sound, Print and Television Recording Films	251
Film Width Standards	252
Film Projection Speed	252
Appendix B: Table of Frequencies	**253**
Frequency Bands	253
Radio Bands	253
Appendix C: Television Channels, Standards, and Systems	**254**
Television Channels	254
Existing Television Picture Standards	254
Proposed Television Picture Standards: HDTV	254
Color Television Systems	254
Appendix D: Film Emulsion Speed Conversion Table	**255**
Appendix E: Television Systems Worldwide	**256**
Appendix F: National and International News Agencies	**261**

Preface

Theory and practice often differ, and reading and learning require practical applications, working knowledge, and hands-on experience. As in medicine, there can be no good training without the benefit of a clinic and a hospital, similarly there can be no good training in broadcasting and film without the benefit of studio and location work. Each requires its own professional jargon, its own professional language.

This book started out as a small brochure—a production-oriented terminology and learning guide to address the practical language of the broadcast and film media. This book aims to fill the need for a professional language and terminology guide to international radio, television, and film production, giving the necessary technical information for the right "call," and the smooth "roll."

Further, it provides information on professional organizations and guilds; awards and award presentations; festivals; national and international standards; film and television systems; the various news agencies aiding the roving reporter, videographer and filmmaker working abroad, the traveling researcher, the student, and the scholar in radio-television broadcasting and film.

Acknowledgments

A book of this nature is an endless task with continual changes, updates and additions. It comes as a result of years of learning, research, work, and experience, conceptualizing and organization, often depending on cooperative efforts and guidance.

In the world of higher learning and academia I must thank Professor Sidney A. Dimond of Boston University, for leading me into the realm of American broadcasting, the late Dr. Robert Steele, who placed film studies into perspective, and Dr. Erik Barnouw, then Head of the Center for Mass Communication, Columbia University, for expanding my cinematic perspectives.

I would like to thank my many colleagues, friends and professional acquaintances both in the United States and abroad. Among them are officials of the Academy of Motion Picture Arts and Sciences; the American Society of Cinematographers; the American Cinema Editors; The American Film Institute, Los Angeles; ASCAP; BMI; ATA—the Association of Radio-Television, Argentina; the British Academy of Film and Television Arts; the British Broadcasting Co., London; Grenada Television, Manchester; and the British Film Institute; ABERT—the Brazilian Broadcasting Co.; Teatro Alfredo Mesquita; Dr. Sabato Magaldi, author/critic in São Paulo; film director Nelson Pereira dos Santos, Zoltan Szmick of UNI Rio, TV-Itapoan, and author Jorge Amado in Brazil; the Canadian Academy of Film and Television; the Canadian Film Institute/Cinémateque Canada; the National Film Board of Canada; dramaturge Dr. Andrew Achim; ORF—the Austrian Broadcasting Company in Vienna, especially Professor Walter Waldherr in Linz, Austria; the Australian Film Institute and Cultural Office; Professor Lifang Zhou, and the Xinhua Institute of Journalism Research, Beijing, People's Republic of China; the Directors Guild of America, Los Angeles; Yvonne von Duehren of the European Film Academy; RTF—French Radio-Television, Paris; Alliance Française, New York City; ZDF—the Second German Television, Mainz/Wiesbaden; Ingrid Scheib-Rothbart of the Goethe House in New York; MRT—Rádio Kossuth; Hungarian Television, Budapest and Pécs; broadcast writer Márta Dálnoky Ráth; and Dr. Ilona Kozma, M&KM, Hungary; the Film Institute of India; International P.E.N.; NHK—Japan Broadcasting Corp. and Mariko Saeki in Tokyo; Voice of Kenya Television; Institute of Mexican

Television; The Motion Picture Association of America, New York; Douglas Barnes, Director of the American Center, Myanmar; the National Association of Television Arts and Sciences, Los Angeles; the Netherlands Broadcasting Co.; author/producer Wim Ibo and film editor James Dunlap, Amsterdam; RAI—Italian Radio & Television, Rome; officers of the National Association of Broadcasters, Washington, D.C.; the Nigerian Television Authority, Lagos, WNTV-Ibadan, Enoh Irukuru Etuk of the Federal Radio Corp., Ambassador A.D. Blankson and Mike Ikenze, Lagos, Nigeria; the New York offices of ITAR-TASS, Russia; the Society of Motion Picture and Television Engineers; Radio Omdurman and Sudan Television, Khartoum, The Sudan; the Writers Guild of America East & West; Television-Zambia and Zambia Educational Television, Lusaka.

I also want to thank the Cultural and Information Offices of the embassies and consulates of Australia, Belgium, Costa Rica, Croatia, Czech Republic, Chile, Denmark, Egypt, Estonia, Finland, Hungary, India, Ireland, Israel, Japan, Lebanon, Lithuania, New Zealand, Norway, Peru, Poland, Portugal, Slovenia, South Korea, Spain, Sweden, Switzerland, and Venezuela.

My sincerest appreciation goes to Michael Roche, who with Greg McCandless and Cindy Boyce of Vermont's Northeast Regional Library scanned the network for all the information I requested; to Lucy Northrop, CPA, for keeping her fingers on my fiscal pulse; to Dr. John J. McKelvey, Jr., and Prof. Edward Chaszar for their friendship and counsel regarding South America and Africa; and to James Gagliardi and Gordon Clow who lent their skills in computer wizardry.

Special thanks go to Frank Lattanzi, who helped with this book in its infancy and read the first version of the manuscript; to Paul R. Beck, Director of Engineering & Technical Manager, Mass Communication, Emerson College, for his technical reading, input, and updates; to Valerie Cimino and Eileen Anderson of Focal Press who handled editorial and production tasks; and to Publishing Director Karen Speerstra for her insight and ever-present friendly encouragement.

I am deeply indebted to Dr. Erik Barnouw, Professor Emeritus and Editor-in-Chief, International Encyclopedia of Communications, who after so many years, again gave me his invaluable suggestions and comments.

Abbreviations

appr. approximately

EC European Community

E/C European/Continental

EE East Europe

e.g. exempli gratia

etc. et cetera

EU European Union

FA French (speaking) Africa

FR France

FRG Federal Republic of Germany

GB Great Britain, United Kingdom of

HQ headquarters

i.e. id est

Inc. incorporated

Ltd. limited

NA North America

U.N. United Nations

US/USA United States of America

vs versus

WHO World Health Organization

International Dictionary
of Broadcasting and Film

A

A   Ampere.

Å   Angstrom.

AA   (1) Agence d'Athenes (Greece). (2) Agencias Andinas (Peru).

AAAA   Also called **4As.** American Association of Advertising Agencies.

AAF   American Advertising Federation.

AANS   Argus African News Service (Zimbabwe).

AAP   Australian Associated Press.

ABA   Alaska Broadcasters Association.

abar   See *microbar.*

ABC   (1) American Broadcasting Company. (2) Australian Broadcasting Commission.

aberration   The distortion of an image introduced by an optical element such as a lens, mirror, or prism.

ABERT   Associação Brasileira de Emissoras de Radio e Televisão (Brazil).

ABES   Association for Broadcast Engineering Standards (USA).

above-the-line cost   Production cost for the story/script, rights, talent, performer, producer, director. See also *below-the-line cost.*

A&B Press   Liberia.

abrasion   Scratches or undesired marks on film surfaces, usually caused by pressure or rubbing against the surface.

A-B roll editing   (electronic A-B roll editing)—two video-audio sources are displayed from two video playback units, processed through a mixer/effects system, and then recorded onto a third recording videotape recorder. Accordingly, assembly programs can be completed from five video/audio or film sources onto a master recorder in A-B-C-D-E roll editing. The process is generally computer-assisted and the sources must use SMPTE/EBU time coding for frame-accurate editing.

A-B rolling   (1) Electronic A-B rolling, television: one film chain contains sound-on-film (SOF) film, while on the second chain silent film is projected. The films can be intermixed (A-B rolled) through the mixing control. (2) In 16mm film editing: all odd numbered shots are put on one reel (A roll) with a black leader replacing the even-numbered shots. The even-numbered shots are then put into another reel (B roll) with a black leader replacing the odd-numbered shots in a checkerboard manner. See also *checkerboard*.

A-B test   Direct comparison of audio, video, data, or RF signals by means of a quality test conducted by switching from one signal to another and monitoring any differences.

ABU   Asia-Pacific Broadcasting Union (Malaysia).

A&B wind   See *A wind, B wind*.

AC   (1) alternating current. (2) assistant camera.

Academia de las Artes y las Ciencias Cinematografias de España
National Academy of Cinematographic Arts and Sciences of Spain. See also *Goya*.

Academy aperture   Also called **Academy mask.** Standard image (frame) size of both 35mm motion picture cameras and projectors since the introduction of sound (SOF), providing space for the sound track. Named after the Academy of Motion Picture Arts and Sciences (USA). Also called *Academy mask* and *movietone frame*.

Academy Awards (the Oscar)   Annual presentation by the Academy of Motion Picture Arts and Sciences with awards given in more than twenty categories (USA).

The first Academy Awards celebration was held on May 16, 1929, at Hollywood's Roosevelt Hotel. Of the fifteen statuettes presented the principal winners were:

> *Wings*—best picture of the year (1927–28)
> Emil Jannings—best actor
> Janet Gaynor—best actress
> Frank Borzaga—best director, *Seventh Heaven*
> Lewis Milestone—best comedy director

Special awards went to Warner Brothers for the pioneer talking picture *The Jazz Singer* and to Charlie Chaplin, producer, director, writer, and star of *The Circus*. See also *Academy of Motion Picture Arts and Sciences* and *Oscar*.

Academy leader   A standard leader attached to the beginning (head) of each release print (also to projection print) with number markings [ranging from 10 (or 8) to 3, each one second apart]. Used for cueing up the film in the projector and for film picture alignment (USA).

Academy mask   See *Academy aperture.*

Academy numbers   See *Academy leader.*

Academy of Canadian Cinema & Television (ACCT)   Academie Canadienne du Cinema et Télévision. Founded in 1979 by the Canadian film and television industry, it is the largest professional organization in the field. Its approximately 2,000 members elect a 35-member board of directors. The Academy's yearly honors include the Genie Awards for excellence in film, the Gemini Awards for English language television and the Prix Gemeux for French language television. ACCT publishes the bi-annual *Who's Who in Canadian Film and Television*, several manuals, and how-to books. See also *Genie Awards, Gemini Awards* and *Prix Gemeaux.*

Academy of Motion Picture Arts and Sciences (AMPAS)   An honorary organization of the motion picture industry, founded in May 1927. Its purpose is to advance the motion picture arts and sciences and to provide incentives for higher levels of technical and professional achievements. (USA). Membership is by invitation only, and its twelve branches today represent actors, art directors, cinematographers, directors, executives, producers, film editors, music, public relations, short films, sound, and writers.

The Academy organizes the annual Academy Award ("Oscar") presentations, maintains the Margaret Herrick Library, the Academy Film Archive, the Samuel Goldwyn Theatre, and conducts educational and cultural activities.

Major publications include the *Academy Players Directory* and the *Annual Index to Motion Picture Credits.* See also *Academy Awards* and *Oscar.*

ACC   Automatic color control.

access time   The hour from 7:00 p.m. to 8:00 p.m. when affiliated stations can broadcast non-network syndicated programs or programs produced locally as per the Prime Time Access Rule.

account   A sales/advertising contract between a client (sponsor, advertising agency) and the broadcast station.

ACCT   Academy of Canadian Cinema & Television.

ACE   American Cinema Editors.

acetate (cellulose acetate)   (1) Also called **cel.** A transparent plastic sheet used in preparation of graphic material. (2) Film base; a slow-burning celluloid material on which a light-sensitive emulsion is applied.

achromatic   Description for an optical element, i.e. a lens, that has been corrected during the manufacturing process for chromatic aberration.

ACI   Agence Congolaise d'Information (Congo).

ACIRTA   Association Catholique Internationale pour la Radio, Télévision et l'Audiovisuel. See *UNDA*.

acoustic feedback   See *feedback*/1.

acoustics   The science of sound; all the influential conditions of a given environment that affect the character of sound produced, recorded, or reproduced in it.

ACP   Agence Camerounaise de Presse (Cameroon).

ACP-C   Agence Centrale de Presse-Communication (France).

ACT   Action for Children's Television (USA).

ACTAT   Association of Cinematography, Television and Allied Technicians (GB).

actinic light   The property of light rays capable of causing chemical changes in a photosensitive emulsion.

action   (1) The process of acting; any movement taking place in front of a microphone, television, and/or film camera. (2) The visual (picture) part of a film (versus sound). (3) Command given by the film director indicating the start of a performance—"action."

active filter   An electronic optical circuit that lowers or raises a portion of the frequency range of a signal or light wave.

active satellite   Communication satellite equipped to receive and retransmit signals. See also *passive satellite*.

actor/actress   Also called **performer** or **talent.** Dramatic performer who plays a part and acts in a program or film.

ACTRA   Alliance of Canadian Cinema, Television and Radio Artists.

ACTS   Advanced Communications Technology Satellite.

actual monitor   See *master monitor*.

actuator   See *jack*/3.

ACTV   Advanced compatible television.

AD   Assistant or associate director.

A/D   Analog-to-digital. See also *D/A*.

adaptation   The dramatization of an existing literary work (novel, story, etc.) rewritten in the form of a script for a broadcast program or film.

adapter   Device mounted to broadcast or film equipment to secure additional and/or different parts to improve their application, i.e. lens adapter.

additive color process   The basis of a color television system where the addition method of the three primary colors, red, green and blue, is used instead of color dyes or pigments. In the motion picture developing process this method is no longer in use. See also *subtractive color process*.

additive printing   Printing method in film processing where a combination of three separate colored sources—red, green and blue— compose the light source that exposes the film. The separation of white light into the three colors (red, green and blue) is achieved with a dichroic mirror.

address track   A special longitudinal track on professional ¾" U-matic videocassettes for SMPTE/EBU time code data. See also *time code*.

ADI   Area of dominant influence.

ad-lib   Ad libitum; unrehearsed, nonscripted speech and/or action used usually to fill time.

ADN   Allgemeiner Deutscher Nachrichtendienst (Germany).

ADORA   Asociacion Dominicana de Radiodifusoras (Dominican Republic).

ADP   Agence Dahoméenne de Presse (Benin).

advance   (1) Sound advance in relation to picture. In film projection, synchronization of the sound and corresponding picture are not parallel with the film, but are arranged several frames apart. Sound is ahead of the picture when correctly synchronized. In 35mm format, the sound advance is 21 frames; in 16mm, 26 frames. (2) Payment in advance of royalties to writers, authors, composers.

Advanced Compatible Television (ACTV)   High-definition television system that enables normal reception on regular, existing receivers. See also *compatible color*.

advertisement   Public announcement of product or commercial goods offered for sale. See also *commercial*.

advertising agency An organization contracted by an advertiser (sponsor) to plan, design, execute, and supervise advertising campaigns or individual advertisements.

advertorial(s) Colloquial for editorial advertising.

A&E Arts and Entertainment cable network (USA).

AEA (1) Actors Equity Association (USA). (2) American Electronics Association.

AECT Association for Educational Communications and Technology (USA).

AER Association of European Radios (Brussels, Belgium).

aerial cinematography Cinematography aimed for a high angle broad view filmed from an airplane, helicopter or a dirigible balloon (blimp) that has been specifically outfitted with a vibration-free camera mount and/or camera lens. Specially developed video and film camera systems also serve this purpose.

aerial image cinematography Also called **visual image photography.** A type of special effects photography whereby a motion picture projector and a camera mounted on an animation stand are driven in sync. The projected image and another actual image is recorded by the camera through a cel and the combined images are recorded on the film.

aeroscope A handheld camera powered by a compressed air motor. Designed in 1912 by Polish scientist Kasimir de Proszybski.

AES Audio Engineering Society (USA).

AF Audio frequency.

af (1) Alternating frequency. (2) Allocated frequency.

AFC Automatic frequency control.

affiliate Independent station (local) having a program contract with a national studio or network. See *primary affiliate* and *secondary affiliate*. See also *independent broadcast station*.

AFFS American Federation of Film Societies.

AFG ITU country code for Afghanistan.

AFI (1) The American Film Institute. (2) Australian Film Institute.

AFI Awards See *The Australian Film Institute Awards*.

AFI Maya Deren Award See *American Film Institute*.

AFM American Federation of Musicians.

AFMA   American Film Marketing Association.

AFN   Armed Forces Network (USA).

AFNOR   Association Francaise de Normalisation (France).

AFP   Agence France Presse (France).

AFRTS   American Forces Radio and Television Service.

AFS   ITU country code for South Africa.

AFT   Automatic fine tuning.

AFTRA   American Federation of Television and Radio Artists.

AG   Authors Guild (USA).

AGC   Automatic gain control.

Agence Belga   Agence Télégraphique Belge de Presse (Belgium).

agency/agent   An organization or individual which, for a fee or percentage, acts, represents, and negotiates on behalf of talent, writers, artists, or a production firm. See also *advertising agency*.

Agfacolor (Agfa-Gaevert)   Negative/positive integral tri-pack color film process made in Germany since 1940.

agitator   (1) A device used in studio water tanks to make waves. It can be operated manually or pneumatically. (2) A device that keeps the developer, stop bath, or fixer in motion during film processing.

AGL   ITU country code for Angola.

AGMA   American Guild of Musical Artists.

AGP   Agence Guiéenne de Presse (Guinea).

AGVA   American Guild of Variety Artists.

AIBD   Asia-Pacific Institute for Broadcasting and Development (Malaysia).

AIG   Agence d'Information Gabonaise (Gabon).

AIME   Association of Information Media and Equipment (USA).

AIP   Agence Ivorienne de Presse (Côte d'Ivoire).

AIR   (1) All India Radio. (2) Asociación Internacional de Radiodifusión (Uruguay). See *IAB*.

AIRC   Association of Independent Radio Companies (GB).

air knife (air squeegee)   Powerful compressed air jet that wipes (blows) off excess liquid from film surfaces during continuous processing. See also *squeegee*.

air quality   See *broadcast quality*.

air time   The scheduled broadcast time of a program.

AIST   Agency of Industrial Science and Technology (Japan).

AKP   Agence Kmere de Presse (Cambodia).

ALA   (1) American Library Association. (2) Authors League of America (since 1912).

ALB   ITU country code for Albania.

A-lens   See *anamorphic lens*.

ALG   ITU country code for Algeria.

alpha loop/wrap   Magnetic tape wound around the drum in a videotape recorder, in the form of the letter α. Rarely used today. See also *omega loop*, and *half loop*.

ALS   ITU country code for Alaska.

alternating current (AC)   Electrical current as supplied by a regular wall outlet: 60Hz at 110–120 volts, and 50Hz at 220–240 volts.

alternating voices   See *multiple voices*.

AM   (1) Ante meridian. (2) Amplitude modulation. (3) Advertising manager.

AMARC   Association Mondial de la Radio Communauté. See *The World Association of Community Radio Broadcasters* (Montreal, Canada).

AMAX   "AM at its maximum"; certification awarded by the National Association of Broadcasters and the Electronic Industries Association to high-quality AM stereo receivers that meet superior technical standards (USA). See *National Association of Broadcasters*.

ambience   Ambient sound. See *room tone*.

American Cinema Editors (ACE)   An honorary professional organization founded in 1950 to give special recognition to film editors of outstanding ability and to advance the art, prestige, and dignity of the film editing profession. Besides the active, inactive, and associate members, there are affiliates, life and honorary members. Membership criteria was broadened in 1970 to include editors from outside the United States. ACE publishes the *American Cinemeditor* quarterly.

American Film and Video Festival of EFLA   Defunct. See *National Educational Film & Video Festival.*

American Film Institute (AFI)   Founded by the National Endowment for the Arts following the National Foundation on the Arts and the Humanities Act of 1965, with the aim of "Preserving the heritage of film and television; identifying and training new talent, and increasing recognition and understanding of the moving image as an art form." The Institute coordinates the National Center for Film and Video Preservation, organizes programs and exhibits, and maintains the Louis B. Mayer Library and the Center for Advanced Film and Television Studies. The Center is responsible at its Conservatory for production training programs and for the *AFI Catalog of Feature Films* and NAMID—the National Moving Image Database.

The Robert M. Bennett Award recognizes achievements in television programming and in 1986 the AFI Maya Deren Award for Independent and Video Artists was established. AFI's most visible presentation is the annual Life Achievement Award.

American Society of Cinematographers (ASC)   Founded in 1919 as a non-profit organization with the motto: "Loyalty, progress and artistry." Membership is by invitation only. The Society stems from two earlier organizations: the Static Club, Hollywood, California (1913) and The Cinema Camera Club, New York (1914).

ASC publishes *American Cinematographer*, a monthly magazine, and periodically issues the *American Cinematographer Manual.*

American Society of Composers, Authors (Lyricists) and (Music) Publishers (ASCAP)   The oldest performing rights organization was founded in New York in 1914. Owned by writers and publishers, it handles disbursement and collection of royalties for its members and assures licensees (music users) compliance with the Copyright Law.

ammeter   Instrument used to measure the electrical current between the load and power supply or current source.

amp   Abbreviation for *ampere.*

AMP   Association of Media Producers (USA).

AMPAS   Academy of Motion Picture Arts and Sciences (USA).

ampere (A or amp)   A unit of electrical current flow; one-tenth the absolute ampère or abampère. Related to voltage and resistance as described by Ohm's law. Named after André Marie Ampère (1775–1836), French mathematician and physicist.

amplifier   An electronic circuit designed to increase the level or amplitude of an electrical signal, and the strength of audio, video, RF, or data signals.

amplitude modulation (AM)   RF broadcasting: the variations of the amplitude power of the carrier wave while the main frequency remains the same. See also *frequency modulation* and *modulation*.

amps   Abbreviation for amperes; watts/volts = amperes. Watts ÷ volts = amps.

AMPTP   Alliance of Motion Picture and Television Producers (USA).

AN   (1) Agencia Nacional (Brazil). (2) All night.

anamorphic lens (A-lens)   A specially designed lens for wide-screen cinematography at the ratio of 2.35:1 that horizontally compresses (squeezes) the picture on the print. To correct that compression it is projected onto the screen by lateral expansion. See also *wide-screen process*.

ANARC   Association of North American Radio Clubs.

anastigmat   Lens corrected optically for the aberration of astigmatism. See *aberration,* also *astigmatism*.

Anatolia   Anatolia Ajansi (Turkey).

ANB   Asahi National Broadcasting Co. (Japan).

anchor   Anchorman/woman; key announcer or newscaster who coordinates a news program.

AND   ITU country code for Andorra.

ANDEBU   Asociacion Nacional de Broadcasters Uruguayos (Uruguay).

ANG   ITU country code for Anguilla.

angle   Point of view; the camera-subject relationship; the direction from which the shooting takes place.

angle of light   The angle between the light/subject and camera/subject axis, referring to both the horizontal and the vertical angle.

angle of view   See *field of view*.

angle shot   Camera technique by which a scene or a subject is shot from different angles, usually for dramatic effect. The shots can be termed "wide," "close," "high," or "low" in relation to a standard (normal) lens-equipped eye-level camera shot.

angstrom (Å)   Unit used to express short wavelength; a hundred-millionth of a centimeter. Named after Swedish physicist Anders Jonas Ångström (1814–1874).

ANI (1) Agence National d'Information (Lebanon). (2) Agencia Nacional de Informaciones (Uruguay).

Anik Domestic satellite of Canada (1972).

ANIM Agence National d'Information Malienne (Mali).

animation Process of shooting a number of different, but sequential cartoon drawings to create the illusion of movement. This illusion is achieved in several ways, by drawing directly on the film, drawing on celluloid sheets to be photographed, filming cutouts or moving puppets, or by mechanically producing movement. Recent developments include computer-generated images—CGI. They are often created directly for videotape or other electronic storage media, thus bypassing the traditional process of film animation.

animation camera A precision-built rugged film camera with frame and footage counters, single-frame capabilities, automatic faders, dissolvers and follow focus, and forward and reverse run. It is mounted vertically on a rigid *animation stand.*

animation stand Also called **benchwork.** A rigidly constructed stand that allows the horizontal and vertical movement of the camera mounted on it. It contains a mounting column, a rotation table, and registration pins, and can be operated either manually or automatically. Its usual height is 8–13 ft. (2.4–4 m).

ann. or anncr. Announcer.

announcement Advertising or other message (2 minutes long or less) carried either during a program or between two programs. See also *spot.*

announcer Broadcast station personnel who read on- and off-the-air announcements, news and news summaries, reports and sports reports, reviews, and commercials. Announcers also introduce programs and act as masters of ceremonies.

announcing booth Also called **continuity studio.** A soundproof small studio where announcements, off-camera announcements, commentaries, voice-overs are read.

ANOP Agencia Noticiosa Portuguesa (Portugal).

ANP Algemeen Nederlands Presbureau (The Netherlands).

ANS (1) Agencia Noticiosa Saporiti (Argentina). (2) American National Standards.

ANSA Agenzia Nazionale Stampa Associata (Italy).

Anscochrome American-made integral tri-pack reversal color film process.

Anscocolor   Integral tri-pack color film process using reversal film. Introduced in the United States in 1941. No longer used.

ANSI   American National Standards Institute.

answer print   The first print of a finished film complete with sound and picture, and printed with correct timing for each shot. See *trial print*.

ANT   ITU country code for Antarctica.

ANTEL   Administracion Nacional de Telecomunicaciones (El Salvador).

ANTELCO   Administracion Nacional de Telecomunicaciones (Paraguay).

antenna polarization   See *polarization*.

anti-halation backing   Also called **anti-halation coating** or **anti-reflective coating**. A dark layer applied on the back of film to absorb light, prevent its reflection from the base, and block halation.

AOR   Atlantic Ocean Region; satellite ocean region configuration.

AP   The Associated Press (USA).

APA   Austria Presse Agentur.

APC   Agence Presse Congolaise (Zaire).

aperture   (1) Lens aperture; an opening on a lens, camera, projector, or printer through which light passes. The amount of the passing light can be controlled by an adjustment iris or by a masked opening. See also *T-stop* and *diaphragm*. (2) The reflective area affected by a satellite dish.

APN   Agentsvo Pocati Novosti (Russia).

apogee   A satellite's furthest point of orbit away from the earth.

APP   Associated Press of Pakistan.

APR   (1) Formerly American Public Radio, now called **PRI—Public Radio International.** (2) Asociacion Panamena de Radiodifusion (Panama). (3) Asia Pacific Region; satellite ocean region configuration.

APRA   Australasian Performing Rights Association, Ltd. (Australia).

APS   (1) Algeria Press Service. (2) Agence de Presse Senegalaise (Senegal).

APV   Agence de Presse Voltaique (Burkina Faso).

AR Agencja Robatnicza (Poland).

Arabsat Arab States Satellite organization (Saudi Arabia).

ARB American Research Bureau—a television rating service. See also *DMA rating.*

arc (1) Type of electric lamp, usually operating on direct current, which produces light by a bridge of electricity between two electrodes. There are carbon arc, mercury arc, and xenon arc lamps. The color temperature of a carbon arc lamp is similar to that of sunlight, making it ideal for daylight location work. (2) Camera dolly movement slightly curved in and/or out.

ARC Automatic recording check; permits a review of image quality in the last second of each segment in a video recorder.

ARCHI Asociacion de Radiodifusoras de Chile.

ARCOM Arctic Communication.

ARD Arbeitsgemeinschaftder Öffentlich-Rechtlichen Rundfunkanstallten der Bundesrepublik Deutschland (Germany). See also *BR, DLF, DW, HR, MDR/2, NDR, ORB, RB/1, SDR, SFB, SSR/1, SWF, WDR.*

ARG ITU country code for Argentina.

Ariane Arianespace European satellite consortium (France).

ARM ITU country code for Armenia.

ARP Asociacion de Radiodifusoras del Peru.

ARPA Asociacion de Radiodifusoras Privadas de Argentina.

arranger Musician/composer who orchestrates musical compositions for radio, television, or motion pictures. The arrangement may be a theme, a full score, or just musical bridges.

ARRL American Radio Relay League.

ARS ITU country code for Saudi Arabia.

art director Art professional, often the head of the art department, who designs and supervises the construction of the sets. See also *production designer.*

ARTE Franco-German cultural television channel.

artificial light Light produced by an electrical or other power source as opposed to natural light or daylight. See also *daylight, natural light,* and *sunlight.*

artist (1) Fine artist, painter, illustrator, or musician who works on and contributes to a broadcast or film program. (2) British term for actor or talent.

ARU ITU country code for Aruba.

ASA American Standards Association.

ASA numbers Internationally accepted standard for rating film emulsion speed, as established by the American Standards Association. Replaced by the ratings of the International Standards Association. See *ISO* and *DIN*.

ASBORA Asociacion Boliviana de Radiodifusion (Bolivia).

ASBU Arab States Broadcasting Union (Egypt).

ASC (1) ITU country code for Ascension Island. (2) American Society of Cinematographers.

ASCAP American Society of Composers, Authors and Publishers.

ASDER Asociacion Salvadoreña de Empresarios de Radiodifusion (El Salvador).

ashcan Colloquial for a 1,000-watt light.

AsiaSat Asia Satellite Telecommunications; the first commercial satellite launched by China in 1990 for a Singapore concern (Hong Kong).

ASIFA See *IAFA*.

aspect ratio Also called **format.** (1) The width-to-height ratio of a standard television picture and/or motion picture frame: four units wide and three units high—4:3, i.e. 1.33:1. (2) HDTV standard of 5:3 (16:9), i.e. 1.78:1. See also *wide screen process.*

assemble edit A type of videotape edit that entails making an edit point where new video, new audio, and new control track pulses are added to a previous segment of similar information. This videotape edit is analogous to the cinematographic *butt splice.*

assembly The initial arrangement of film shots and scenes following the scripted order.

assignment A specific task, usually referring to news coverage and/or background research to a story given (assigned) to a reporter or news or production team.

assignment editor News (city desk) editor at a broadcast station who assigns reporters or news team(s) to cover specific (news) events.

assignment of copyright See *transfer/1.*

assistant cameraman/woman   See *camera assistant.*

assistant director (AD)   The director's first assistant for coordinating all production activities in the creative process. May also act as liaison with authorities, officials, and supply houses.

assistant producer   The producer's right hand man/woman, often charged with supervisory functions involving preproduction arrangements, and administrative and production duties.

associate producer   The producer's associate or partner in all business, organizational, and/or creative matters of a broadcast/film production.

astigmatism   Optical defect in a lens that prevents rays of light from being brought to common focus. Better quality lenses are corrected for astigmatism. See *anastigmat.*

ASTRA   The first privately owned European communication satellite launched in 1988 by the Société Européenne des Satellites (Luxembourg).

ASTVC   American Society of Television Cameramen/women.

ATA   (1) Agence Télégraphique Albanaise (Albania). (2) Asociacion de Teleradiodifusoras Argentinas.

ATG   ITU country code for Antigua.

atmosphere   Artistic setting, milieu; specific lighting, any action or personnel appearing in the background of a scene.

atmospheric distortion   Abnormal broadcast signal propagation and disturbances due usually to weather changes and conditions.

ATN   ITU country code for The Netherlands Antilles.

ATP   Agence Tobadienne de Presse (Chad).

ATR   Audio tape recorder.

ATS   Agence Télégraphique Suisse/Schweitzerische Depeschenagentur—SDA (Switzerland).

ATSC   Advanced Television Systems Committee (USA).

ATTC   Advanced Television Test Center (USA).

attenuation   (1) The reduction of signal strength while it travels through cable. The higher the frequency, the higher the rate of attenuation. (2) The gradual loss of energy of the traveling energy wave.

ATV   Advanced television.

audience   The listener or viewer.

audience participation   Broadcast program, like a talk or game show, where the audience participates. See also *quiz program.*

audience rating   See *rating*/1.

audio   (1) Audible sound as reproduced electronically; radio frequency or power circuits. Used also to designate audible frequencies, the normal being 50–15,000 Hz. (20–20,000 Hz are extreme limits). See also *frequency.* (2) The accompanying sound to the visual image in television transmission. (3) The sound portion of a motion picture film.

audio amplifier   See *amplifier.*

audio band   See *band*/1.

audio cue   See *cue*/1.

audio engineer   See *sound engineer.*

audio head   See *head*/2.

audio mixer   See *mixer*/2.

audio signal   See *signal.*

audio tape   See *magnetic tape.*

audio tape recorder (ATR)   An electrical device used to record and/or reproduce audio signals, i.e. sound. See also *videotape recorder.*

audiovisual (audiovisuals, AV)   Presentation technique using sound and vision—sound tapes, records, still photographs, slides, film, and filmstrips—in a coordinated manner. Used increasingly in education, industrial training, business, marketing and sales presentations.

audition   (1) Trial hearing or testing of a performer's (talent's) abilities, often in front of a microphone, television, or film camera and judged by the studio and station (and/or agency) personnel. See also *casting* and *screen test.* (2) Audition channel; second channel in an audio mixing console, used specifically to "pre-hear" audio sources prior to sending them out as program or broadcast signals.

AUP   Australian United Press.

AUS   ITU country code for Australia.

Aussat   A general purpose Australian satellite.

Australian Film Institute (AFI)   A non-profit organization founded in 1958 to increase knowledge, appreciation, and enjoyment of the art of film. Has a membership of over 8,000, and is open to film and television professionals as well as the interested public.

AFI operates the Cinesearch research service and maintains a specialist film literature collection. The major distributor of Australian independent films, AFI publishes *Cinedossier*, a monthly media review of film and television, and organizes the annual Australian Film Festival.

Australian Film Institute Awards, The The AFI Awards, Australia's most prestigious, were established in 1958 with the aims of recognizing achievement, encouraging excellence in Australian film making and promoting awareness of Australian film production. The categories are in feature films, short fiction, animation, and documentary. The special Raymond Longford Award and the Byron Kennedy Award are given to outstanding individuals, and another special award is presented for the Best Foreign Film. Since 1986 television drama and television documentary have been included and honored in the annual event.

AUT ITU country code for Austria.

auteur Author.

authorization See *copyright, model release,* and *property release.*

author of the script See *scriptwriter.*

author/theory Critical theory that considers the director the real "author" of a film.

autocue See *prompter.*

automatic frequency control (AFC) A receiver circuit to prevent frequency drifting to help it stay "on channel."

automatic gain control (AGC) A device in the sound or video recorder or RF receiver that automatically maintains a steady signal level.

automatic shut-off Also called **end of tape—EOT.** (1) A special switching mechanism on a tape recorder that stops the motor automatically when the tape ends or breaks. (2) A device built into a film camera that automatically triggers/shuts off the camera motor when the film breaks or runs out. See also *buckle switch.*

automatic slate See *beep.*

A/V Audiovisual(s).

available light See *existing light.*

avant garde film Innovative, unorthodox film in concept as well as in technique.

AVC   Association of Visual Communicators, formerly IFPA (USA).

AVD   Alternating voice and data (circuits).

AVI   Agence Vietnamienne d'Information (Vietnam).

AVP   Audiovisual production division of UNESCO. See *United Nations Educational, Scientific and Cultural Organization.*

AVR   Audiovisual recorder.

AWG   Australian Writers Guild.

A-wind   Reel of single perforated film unwinding clockwise with the emulsion coating (dull side) facing toward the hub or inside of the reel with the perforations toward the camera operator. Films with A-wind are used for making contact prints. Refers mostly to 16mm film stock. See also *B-wind.*

AWR   Adventist World Radio (USA).

AXIS   See *projection axis.*

AZE   ITU country code for Azerbaijan.

Az/El mount   Azimuth/elevation mount. See *mount/2.*

azimuth adjustment   Adjustment of the head gap to a position perpendicular to the horizontal base of the tape in a recorder.

AZR   ITU country code for Azores.

B

B (1) ITU country code for Brazil. (2) Basic (cable).

babble Interference; disturbing sound; crosstalk in the telecommunication system.

baby Also called **baby keg** or **baby spot.** Colloquial term for a small spotlight with a 500- or 750-watt bulb housed in a casing, usually equipped with a lens.

baby legs Baby tripod; a short-legged tripod used for filming at a low angle.

backdrop See *drop*.

background (background level) (1) Sound effects, dialogue or music at a low level for background effect. "Hold to BG" means to turn the volume down and hold under as background. "Hold under" direction is also used. See also *under*. (2) Part of the scene or picture that serves as setting for the action.

background level See *under*.

background light Also called **dressing light** or **set light.** Light illuminating the background (scenery). See also *back-light, fill light* and *key light*.

background music Music other than the main theme used in a broadcast program or film for mood and underlying effect.

background plate A print produced especially for process photography, e.g. background projection.

backing See *anti-halation backing*.

back laying A film editing room technique to synchronize the end, rather than the start, of a scene with the sound track.

back-light/lamp Also called **separation light.** Illumination directed from behind and above the subject, opposite the camera, to separate the subject from the background to create a three-dimensional effect. See also *background light, fill light* and *key light*.

back-light (fluorescence) ultraviolet photography   See *ultraviolet photography*.

back lot   See *lot*.

back projection   (1) A projection method whereby the projector is placed behind a translucent screen and the picture is viewed from the opposite side. This method, although becoming obsolete, is used in commercial presentations because it does not require a darkened room. See also *front projection*. (2) A specific scene that is projected on a translucent screen to serve as background for a performance. Both the action and the background scene are photographed by the camera, thus simulating "location" filming in the studio (sound stage). See *traveling matte*.

back timing   The backwards timing of a broadcast program from its end to the beginning. Back timing is done to help the director and performer pace (time) the show correctly, enabling them to finish "on the nose" as scheduled. Important in radio and television.

back-up copy   A high quality videotape or film copy made for protection.

back-up track   A sound track used as "security" in editing. See also *scratch track*.

bad footage   Also called **N.G.** Film footage rendered unsatisfactory.

bad take   Unsatisfactory recording or shot; N.G.

BAEA   British Actors Equity Association.

baffle   (1) Panel or front panel where most speakers are mounted (2) Louvered shutter in front of a studio lamp for light direction and density control. (3) Sound baffle: a blanket (usually of textile material) to absorb and muffle sound.

BAFTA   British Academy of Film and Television Arts.

BAH   ITU country code for the Bahamas.

balance   (1) Audio: A harmonious, balanced blend of the various sounds in production. The mixing of sound to create a natural effect. (2) Video: An even, satisfactory picture composition on the screen. See *white level*.

balance head   A tripod head with a movable tie-down (screw) for television cameras and a heavy-duty spring for tension adjustment to counterbalance the weight of long lenses.

balance stripe   Magnetic stripe on the film, opposite the edge of the magnetic sound track. See also *magnetic track*.

Balázs, Béla Award   Yearly award presentation in ten categories given by the Ministry of Education and Culture of the Republic of Hungary for outstanding contributions in motion pictures and television.

ball leveling   Camera tripod head that can be adjusted to a level position by a movable ball that can be locked beneath the base.

balop   (1) Balopticon, an opaque television projector, now of historical value. (2) Slide (opaque or transparent) used in the balopticon.

Balticum Film & TV Festival   Documentary film and television program festival for Baltic countries: Denmark, Sweden, Finland, Russia, Estonia, Latvia, Lithuania, Poland, and Germany (Denmark).

BAM   Broadcast Authority of Malta.

Banco Nacional International Film Festival   Non-competitive annual festival for American and European entertainment and feature films, sponsored by the National Bank and held in Rio de Janeiro (Brazil).

band   (1) Frequency band; the range of frequencies within two definite limits. The standard radio broadcast band extends from 530 to 1600 kHz; television bands—VHF and UHF—extend from 41 to 850 MHz, and are divided into five ITU regulated groups. (2) An individual selection on a transcription. See *cut/2*

bandwidth (BW)   Also called **channel width.** The limit of the frequency spectrum, expressed in Hz (kHz or MHz), assigned to a specific channel (station).

bank   An arrangement or row of lamps in one large casing.

bar   Unit of air pressure, equal to 1,000,000 dynes per square cm (approximately 14.50 psi)—somewhat less than one standard atmosphere. In acoustics, a unit formerly called bar equals one dyne per square cm. It is now called **microbar** or **barye.**

barn doors   Metal doors (flaps) hinged to the front of studio lamps to control the direction of the light beam and/or to shade subject areas.

barney   Flexible, insulated cover over a film camera used to reduce camera noise in a double-system, synchronized sound filming. It is a substitute for a *blimp.*

barrel   See *lens barrel.*

barye   See *microbar.*

base   (1) (base light) The even, general studio illumination required for setting television cameras to achieve normal camera pick-up. See also *modeling light*. (2) Film base or tape base; a flexible (cellulose) material on which light-sensitive film emulsion or oxide layer is applied during manufacturing. See also *motion picture film* and *magnetic tape*.

base band   Unmodulated (non-modulated), basic analog audio and video signal, used in decoders.

basher   See *camera light*.

basic cable   (1) Basic cable network; broadcast network system that receives payments for programs from cable networks that carry it and sells commercials to its own clients, leaving time hourly for local advertisements. (2) Lowest cost for consumer cable subscriber; lowest tier of basic service.

BATELCO   The Bahamas Telecommunication Corporation.

battery   Also called **battery pack** and **power pack.** Portable, usually rechargeable, battery (pack), supplying DC power to camera or recorder. According to their use and placement there may be internal, belt, shoulder pack, trolley (wheeled) and on-board (snap-on) batteries. See also *Nicad*.

battery belt   A series of lightweight batteries mounted in a wide (leather) belt, similar to an ammunition belt, and worn by the cameraman/woman to facilitate movement, especially in handheld operation. It supplies voltage from 6 to 30 V.

battery cable   See *power cable*.

battery charger   Various types of compact units that plug into a regular wall outlet and charge batteries in a few hours or overnight.

bayonet mount   A lens mount where prongs (bayonets) on the base of a lens fit into the slots on the camera to facilitate lens changes. See also *C-mount*.

BBC   British Broadcasting Corporation.

BBCWS   BBC World Service.

BBS   Bhutan Broadcasting Service.

BC (B'cast)   Abbreviation for *broadcast*.

BCB   Broadcast band; refers to medium wave band.

BCC   Broadcasting Corporation of China (Republic).

BCFM   Broadcast Cable Financial Management Association (USA).

BDA   Broadcast Designers Association (USA).

BDI   ITU country code for Burundi.

BEA   Broadcast Education Association (USA).

beam   (1) Electron beam. (2) CB slang for directional antenna.

beard   An over-the-air error, a fluff, made by the announcer.

beep   A short, high-pitched audio signal used as a cue mark. In double-system (sync-sound) film production, modern cameras are equipped to automatically trigger a "beep" sound with a corresponding light "pin" that marks (exposes) a small dot on the film.

behind-the-lens (BTL)   Through-the-lens (TTL) viewing or metering device.

Beijing International Scientific Film Festival   Yearly festival for films 60 minutes or less in science, technology, environment, nature, and medicine (China).

bel   Ten decibel unit expressing comparison of two levels of power in an electrical communication circuit. Named after Alexander Graham Bell (1847–1922), American inventor.

BEL   ITU country code for Belgium.

Bell & Howell Perforation   See *negative perforation*.

bellows   A light-proof, treated cloth device, folded like an accordion, with metal holders in a square shape and a lens mount on one end, attached to a film camera to improve the focal length of the lens used.

below-the-line cost   Production cost for the technical crew, production/unit manager, assistant director(s), script supervisor, make-up, equipment and studio rental, art department, sets, location fees, housing and transportation, raw stock, editing, titles, effects, laboratory fees, insurance and taxes. See also *above-the-line cost*.

belt battery   See *battery belt*.

Belva Brissett Award   See *National Association of Broadcasters*.

BEN   ITU country code for Benin.

benchwork   See *animation stand*.

Berlin Bear   Award of the International Film Festival held in Berlin, Germany.

best boy   First assistant electrician.

Best of the Best Award   See *National Association of Broadcasters*.

Betacam   One-half-inch-wide broadcast quality video component recording format, a Sony development, with video cassettes coming in 3 lengths—30, 60, or 90 minutes.

Betamax   One-half-inch videotape format housed in a small cassette for home consumer and industrial applications (Betamax, Beta Hi-Fi, Super Beta, Ed Beta).

BFA   (1) ITU country code for Burkina Faso. (2) British Film Academy. (3) Broadcast Foundation of America.

BFBS   British Forces Broadcasting Service.

BFI   British Film Institute.

BG   Abbreviation for *background*.

BGD   ITU country code for Bangladesh.

BH   Bell & Howell. See also *KS*.

BHR   ITU country code for Bahrain.

BH standard   See *negative perforation*.

BHU   ITU country code for Bhutan.

bias   (1) A high-frequency alternating current fed into a recording circuit to eliminate distortion and facilitate the magnetic recording process. (2) A steady voltage applied to an element in an electrical circuit, causing it to take on a different from normal form.

BIC   Broadcast Industry Council (USA).

bicycling   Videotape or filmed program exchange between broadcast stations to fill delayed or last minute broadcast slots.

bi-directional mike   Microphone with a pick-up pattern in a figure eight shape. See also *omni-directional mike* and *uni-directional mike*.

Biennale di Venezia, La; Mostra Internazionale d'Arte Cinematografica   Bi-annual international Film Festival of Venice with programs in several categories, including theatrical features, art films, children's films and documentaries. The Film Festival is part of the world's oldest (1895) international art exhibition (Italy).

Big Mac   10-kW controlled Fresnel spotlight.

BIH   International country code for Bosnia-Herzegovina.

billboards   Short listings of the opening and closing of broadcast programs giving the names of participating advertisers.

bin   See *trim bin*.

binaural   Sound transmitted from two sources. Its tone and pitch may vary acoustically. See also *monaural* and *quadraphonic*.

binaural jack   Tape recorder output jack that accepts binaural earphones.

binaural sound   Sound recorded on two channels, each heard with one ear; i.e. left ear: channel 1, right ear: channel 2.

bioscope   An early projection device developed by Skladanowski in Germany and based on the Edison principle.

BIP   Brussels International Press Agency (Belgium).

bi-pack   A method whereby two films are run simultaneously through the camera. Often used in special effects filming.

bird   Colloquial for satellite.

bird's eye view   See *high-angle shot*.

bit part   A small speaking part in a show or film.

bit player   Actor or talent playing a bit part.

BJ'er   See *broadcast journalist*.

BKSTS   British Kinematograph Sound and Television Society.

black   (1) The relative darkest point of the video screen having no illumination. (2) The darkest part of the gray scale (marked No. 10). (3) Video black—black with set-up at 7.5 to 10 IRE. See also *black level*. (4) Crystal black—set up at 0 IRE. (5) "To black"—a command by the television director to indicate a fade to black or fade out.

black and white (B&W)   (1) Monochrome television. (2) Monochrome motion picture film.

black jack   Actuator. See *jack/3*.

black leader   Black film strip (leader) used in editing and/or film laboratories, especially for producing A&B rolls.

black level   Also called **pedestal.** Black balance; 7.5% above zero illumination or blackness in the transmitted video signal. See *black/3*. See also *white level*.

black light   Near-ultraviolet light; radiant energy that falls just outside the visible spectrum. Used with filters for special effects in both B&W and color filming.

Black Maria   The name of Thomas Alva Edison's first studio, as he liked to call it (USA).

black net   Heavy, light-absorbing open-end net. See also *open-end net*.

black reference   See *reference black*.

blank   (1) A piece of tape or film without sound or images. (2) See *freeze*/2.

blanket license   Musical license granting broadcast stations, studios, and public places the right to use of music described in a list or repertoire for an agreed-upon period and for a set fee.

blanking interval   Part of a video signal reserved for special synchronizing and test signals. Also used frequently for captioning and teletext data.

blast   Distorted, loud sound caused by high volume levels. In modern broadcasting, volume control is regulated automatically.

bleep tone   See *beep*.

blimp   A lightweight (fiber or aluminum alloy) cover in which the film camera is located during sound filming to prevent camera noise from being picked up by the microphone. The blimp is equipped with exterior camera and lens controls. See also *barney*.

BlimpCam   Blimped camera.

block   A series of similar broadcast programs.

blocking   The working process by which a director arranges the action and movements of the performers, positions the cameras, microphones, and lighting, and generally sets the pace and action of a show. Usually done during rehearsals.

bloom   (1) A bright illumination of a picture or part of it on the picture tube that obscures picture detail, resulting in flare, ghost images, and diminishing picture contrast. (2) British term for anti-reflection coating on a lens.

bloop   A splice on the sound tape causing an intentional or unintentional dull sound or an unwanted noise.

blow-up   (1) Enlargement of a photograph, or part thereof, or other printed material for effective television transmission. (2) Laboratory technique, using an optical printer, by which a smaller-gauge film can be enlarged (blown-up) to a larger format to attain better projection; e.g. 16mm to 35mm.

BLR   ITU country code for Belarus.

blue   (1) Additive primary color. (2) Blue conversion filter (80A) that transforms artificial (tungsten) light into daylight. See also *daylight conversion* and *orange filter*.

blue net   Medium density, open-end net. See also *open-end net.*

Blue Ribbon Award   Film award given to the best non-feature (short and documentary subject) film at the yearly American Film Festival. Obsolete. See *Educational Film Library Association.*

blue screen   See *traveling matte.*

blur   A confused, disfigured and dimmed image on the screen caused at times by a swish pan with the camera.

blurb   Originally, a publisher's synopsis of a book, or same printed on the book jacket. In broadcasting—a publicity release.

blurred picture   An out-of-focus image.

BLZ   ITU country code for Belize.

BMI   Broadcast Music Incorporated (USA).

BNS   Baltic News Service (Lithuania).

board   The control panel or mixer in the control room. See *console.*

body brace   A portable camera support attached to a camera operator's shoulder and waist to facilitate camera movement. See also *shoulder pod* and *Steadicam.*™

body double   A substitute actor or actress for nude (illicit) scenes. See also *double.*

BOL   ITU country code for Bolivia.

boob tube   Colloquial for television set.

book   (1) A two-fold flat part of the scenery. (2) The script. (3) "To book"—booking; to arrange, to contract actor(s), production crew, equipment, studio, transportation, and accommodations.

boom   (1) An extendable arm on which a microphone is mounted and rotated to different directions and height for better sound pick-up. (2) A mobile camera mount, usually on a perambulator, that can be moved in all directions and raised above the set or scenery, enabling the camera to photograph from various angles.

boost   The amplification of sound or electrical frequencies to a desired level.

booster   (1) An electrical device that helps to increase the wattage and the light output of a lamp. (2) An auxiliary arc light used in daylight (location) filming, designed to increase illumination and act as filler light to improve shadow detail.

booth   See *announcing booth.*

boresight viewer   A viewing device that is attached to the side of a film camera. Viewing is achieved by a pinhole through the lens. Focusing is adjusted and then the viewer is replaced by a light-tight plate during filming. Used in special (high speed) cameras.

BOT   ITU country code for Botswana.

bounced light   Light reflected from a ceiling or other large light (white) surfaces.

box office   The cashier's booth of a movie theater where tickets are sold.

B picture   A second rate, usually inexpensive theatrical film.

BPME   Broadcast Promotion and Marketing Executives (USA).

BR   (1) Bayerische Rundfunk (Germany). (2) Berliner Rundfunk (Germany).

braces   Stage braces for holding up scenery.

BRB   ITU country code for Barbados.

BRC   Broadcast Rating Council (USA).

break   (1) A pause in rehearsal; time out. (2) Station break; a pause in the program for station identification and occasionally other announcements.

breakdown   (1) Script breakdown; a schedule that specifies actors, locations, and props that will be needed for each set-up. (2) In film editing, it refers mainly to dailies and is a listing of shots organized in continuity.

B.R.F.   Belgisches Rundfunk und Fernsehzentrum (Belgium).

bridge   A musical and/or visual transition to link scenes and episodes.

bridging   A longer broadcast program scheduled to overlap the starting time of another, usually competing, show.

bridging connection   A non-loading or high impedance connection for audio, video, or data.

brightness   See *luminance*.

brightness control   See *luminance channel*.

brilliance   See *luminance*.

"bring to focus"   See *focusing*.

British Academy Awards, The   Annual awards presentations by the British Academy of Film and Television Arts in various categories for both film and television. (The British Academy Award is based on a design by Mitzi Cunliff.) See also *The Lloyds Bank BAFTA Awards*.

British Academy of Film and Television Arts, The (BAFTA)   The British Film Academy (1947) and The Guild of Television Producers and Directors (1953) formed the Society of Film and Television Arts, which in 1957 became The British Academy of Film and Television Arts. It has an elected membership of approximately 3,000 who are contributors to the film and television industry in Great Britain. BAFTA organizes the annual British Academy Awards presentations, The Lloyds Bank BAFTA Performance Awards, including the Alexander Korda Award, the David Lean Award, and the Flaherty Documentary Award; BAFTA also maintains the Queen Ann Theatre, and the smaller Run Run Shaw Theatre, and conducts various functions in the film and television arts.

British Film Academy, The   The former organization founded in 1947 by a group of film makers that included David Lean, Michael Balcon, and Lawrence Olivier. In 1957 it merged with The Guild of Television Producers and Directors to form The British Academy of Film and Television Arts. See *British Academy of Film and Television Arts*.

British Film Institute (BFI)   The British Film Institute was established in 1933 to encourage the development of film as a work of art, and to promote its use and appreciation, including films for television. The Institute maintains the National Film and Television Archive, BFI Library and Information Services, BFI Research, the National Film Theatre, and the Museum of the Moving Image (MOMI). It organizes exhibitions and distributions, and the London Film Festival. BFI publishes *Sight and Sound* magazine, and various books and series.

BRM   ITU country code for Myanmar (Burma).

broad   Broad light; a large floodlight, used mostly in color television lighting or in color filming to build up the general level of illumination.

broadcast   The transmission of radio and/or television programs by airwaves or cables to individual receivers.

broadcast band   See *band*.

broadcast basic (BB)   Cable television service designation for television station and public access station channels 2 through 13.

Broadcasting Hall of Fame   See *National Association of Broadcasters*.

broadcast journalism   A term stemming from World War II when radio became an important news medium, and in the fifties expanded to include television.

broadcast journalist   Also called **BJ'er**. Radio or television news reporter, correspondent, newscaster, anchor, or editor.

Broadcast Music Incorporated (BMI)   Founded in New York in 1940 to protect the performing rights of composers, songwriters, and music publishers and to license, collect, and distribute royalties. Musical categories include pop, country, rhythm & blues, rock, jazz, Latin, gospel and contemporary concert music (USA).
    BMI publications include the *Handbook for Songwriters & Publishers, Songwriters & Copyrights, A Guide to Music Publishing Terminology,* and *BMI Music World* quarterly.

broadcast quality   Also called **professional quality.** Video or tape recording of the highest (transmittable) quality and stability. See also *consumer grade* and *industrial grade*.

broadcast spectrum   Portion of the frequency range assigned to radio or television stations.

broadcast teletext   Information transmitted on the spare lines of conventional television signal to the home receiver. See *CEEFAX* and *Viewdata*.

BRTN   Belgische Radio en Televisie Nederlandse Uitzendingen (Belgium). See also *RTBF*.

BRU   ITU country code for Brunei.

brute   High-intensity, 225-amp carbon arc lamp. See also *mini brute*.

BS   British Standards.

BSB   British Satellite Broadcasting.

BSC   (1) British Society of Cinematographers. (2) Broadcasting Standards Council (GB).

BSI   British Standards Institute.

BSkyB   British Sky Broadcasting (uses ASTRA satellite).

BTA   (1) Broadcast Technology Associates (USA). (2) Bulgarska Telegrafna Agencia (Bulgaria).

BTL   Behind-the-lens (metering). See also *TTL*.

BTX   Bildschirmtext; videotext terminal (Germany).

bucket   Colloquial term for floodlight.

buckle Jam; film jammed in the camera. See also *spaghetti*.

buckle switch Run out switch; a trip switch, incorporated in film cameras (especially high-speed cameras) to break the power current in case of a "buckle," film breakage, pile up, or threading failure.

buckling Shrinkage of the outside edge of a roll of film, caused by dry weather conditions. See also *fluting*.

bug Hard-to-trace trouble occurring intermittently in equipment.

build-up See *slug/2*.

BUL ITU country code for Bulgaria.

bulb Specially designed lamp (glass tube) used in lights, projectors, printers, and other equipment.

bulk eraser Electromagnetic device used to bulk eliminate magnetic patterns from video and audio tapes. See also *eraser*.

bull horn See *megaphone*.

bull switch Studio light switch.

bumpers Bridging (transition) material between actual broadcast programs and commercials. Usually prerecorded.

burn-in (1) A negative image retained by the television camera tube that results from continuous focusing on one high-contrast subject, or from the incorrect setting of brightness and contrast controls. Thus the image is retained, burnt-in, even while the camera is picking up another subject. (2) Burning in a part of a photographic image by giving greater exposure to certain area or areas, thus making that area darker.

bus (buss) The rows of buttons on the video control panel.

busy A picture in which the set or background is too cluttered, with too much detail, distracting attention from the main action.

butterfly A scrim or a piece of silk material used to soften shadows and diffuse strong light.

butt splice Splicing the audio tape or film pieces together on the lines of abutment, with no overlapping, fastened across by a transparent tape. See also *lap splice*.

buzzard A bad photographic take.

buzz track Background sound effects track to carry room tone.

B&W Black and white, monochrome.

BW   Bandwidth.

B-wind   In film rolls where only one edge is perforated (single perforation), the emulsion faces inward toward the hub, while the perforations are away from the camera operator. B-wind films are used for camera film, for making opticals, and in bi-directional printers. Refers mostly to 16mm film stock. See also *A-wind*.

CAB Canadian Association of Broadcasters.

cable See *camera cable, coaxial cable, mike cable* and *power cable.*

cable guard A metal guard around the wheels of a tripod dolly or pedestal preventing the cable from rolling under the wheels.

cable release A film camera release consisting of a flexible sheath with an inner cable, used when camera stability is critical, i.e. single-frame exposure.

cable TV (cable television) Television programs carried by a coaxial cable versus airwave transmission. Coaxial cable can carry a wide band of frequencies. See also *central aerial television.*

CAF (1) ITU country code for the Central African Republic. (2) Centralna Agencja Fotograficzna (Poland).

Cairo International Film Festival Annual film festival in various categories and awarding the Golden Nile and Nile Prizes (Egypt).

Calans Caribbean and Latin American News Service (Puerto Rico).

calibration (1) Markings on a lens housing indicating f-stop (f-number) or T-stop (T-number) positions and the distance scale for focusing. (2) Markings found on audio, video and RF instruments.

call (1) Time schedule for summoning actors and production personnel. (2) A casting offer for a role or part in a broadcast program or film. (3) Call for action.

call letters/numbers The initials and numbers assigned by the international communication authorities (Federal Communications Commission, Ministry of Information or Communications, etc.), in conjunction with international agreement, for identification of broadcast stations. Applicants usually try to obtain appropriate call letters (ABC, CBS, CNN, NBC, BBC, CBC, etc.). See also *K/2* and *W/2.*

call sheet A listing of the time, dates, hours for technical crew and performers to report for rehearsal or production.

call sign Colloquial term for call numbers and letters. See *call letters/numbers*.

call time See *call/1*.

cam A revolving device in the film camera or film projector that activates the claws in an intermittent movement.

camcorder Abbreviation for video camera/recorder.

cameo A lighting technique where only the performers are lighted against a dark background.

cameo appearance A short appearance of a popular, well-known personality in a broadcast program or film.

camera (1) Television: (a) part of the television system, hooked up by cables, that picks up the images and transforms them into electrical impulses for direct live transmission or for videotape recording. It consists of a camera housing, a camera tube, or solid state image sensor, view-finder, lens or lenses, controls and accessory equipment. (b) A portable, hand-held video camera with a battery pack, used on location, away from the studio, relaying signals by microwaves or via satellite dish. Its application is increasing in *EJ—electronic journalism, ENG—electronic news gathering, ESG—electronic sports gathering,* and *SNG—satellite news gathering.* (2) Motion picture/film: an apparatus that photographs a series of pictures and records them on a continuous strip of film material. It consists of a lightproof camera body, motor, viewfinder, lens or lenses, controls, and a lightproof, built-in or attached magazine to hold the film.

camera and event synchronizer A device in high-speed (HS) cameras designed to pre-start, simultaneously start, or post-start the high-speed camera in relation to the actual event being studied or filmed.

camera angle See *angle*.

camera assistant (AC) Member of the camera crew, preferably a skilled technician, who carries out a variety of tasks around the camera. He/she must be familiar with every aspect of the camera, lenses, types of films and accessories, and must be able to work fast even in stressful situations.

camera buckle See *buckle*.

camera cable (1) Television: connecting cable between camera and control unit. (2) Motion picture/film: cable connecting the film camera and the battery or other power current.

camera car A specially designed vehicle with various types of camera mounts and/or platforms to facilitate moving shots. See also *shot-maker.*

camera card See *shot sheet.*

camera chain Also called a **camera system.** Television camera and connected equipment such as power supply, cables, sync generator, control units.

camera coverage See *field of vision.*

camera crane See *crane.*

camera crew The camera unit—director of photography, cameraman/woman, camera assistant(s), gaffer and his/her crew.

camera dolly See *dolly.*

Camera d'Or Golden Camera; award of the Cannes International Film Festival, given for cinematography (France). See *Festival International du Film—Cannes.*

camera film Film stock for the camera, used to photograph the actual (original) scene. See also *laboratory film* and *print film.*

camera helmet A small, remote-controlled camera mounted on a helmet of a skier, motorcyclist, car racer, parachutist, or stunt performer to realistically capture the racer/driver's point of view. See also *point-of-view camera.*

camera left/right Command or direction given from the camera's (operator's) point of view, indicating left or right of the camera. "Move camera left," or "camera right." (Opposite of stage left and stage right). See also *stage left/right.*

camera light Also called **basher.** A small, auxiliary spotlight mounted in front of the camera, used to highlight detail. (Not to be confused with the red *tally* or *cue* light on television cameras.)

camera line-up The lining up or set-up of a television or film camera to ready it for shooting. See also *set-up.*

cameraman/woman Member (or head of a camera unit) of a production crew responsible for handling the camera, lighting, and recording of the scene and performance during production. Since both television and motion picture films are termed "visual media," a great deal depends on the cameraman/woman's creative and technical abilities. See also *director of photography.*

camera mount Device to secure and hold video or film camera.

camera movement   The movement of the camera. See *arc, dolly, pan, tilt,* and *trucking.*

camera noise   Audible noise produced by a non-blimped running film camera.

camera obscura   A box painted black on the inside with a small hole on one side through which images are cast that appear inverted on the opposite wall. It was used for viewing eclipses of the sun and for tracing drawings. Now of historical value.

camera operator   Member of the camera crew (responsible to the director of photography) who operates the film camera.

camera original   See *original.*

camera rehearsal   Final rehearsal, similar to a stage dress rehearsal, where the full cast performs in costume with cameras, microphones, sets, props, and lighting before the actual live telecast or videotape recording.

camera report   Also called **log sheet** or **report sheet.** A report form filled out during filming indicating the various scenes, takes, their length, and instructions to the processing laboratory as to what takes (versions) to print. A copy of the camera report is sent to the film editing room.

camera right   See *camera left/right.*

camera shot   See *shot.*

camera slate   See *slate.*

camera speed   The running speed of the film camera expressed in frames per second (fps); the rate by which the film passes through the gate. In the case of a variable speed camera motor (wild motor), the speed can be checked on the tachometer and re-adjusted in the rheostat.

camera system   See *camera chain.*

camera tape   Gaffer tape.

camera truck   A self-contained large truck with a secure, enclosed top to house and carry camera equipment and accessories. It may also have a small maintenance enclosure. Used mostly on location work.

camera tube   Old style, photo-emissive or photo-conductive television tube that receives optical images through the lens and transforms them into electrical impulses. See *iconoscope, Image Orthicon, Vidicon, Plumbicon* and *Saticon.* See also *charged coupled device.*

CAMPP   Canadian Association of Motion Picture Producers.

can   (1) Ashcan. (2) See *film can.*

CAN   ITU country code for Canada.

Canadian Film Institute (CFI)   Institut Canadien du Film—ICF. A federally chartered non-profit organization incorporated in 1935. Governed by a 12-member Board of Directors, the Institute's aim is to promote and encourage the study, appreciation, production, and dissemination of cinematic art for cultural and educational purposes in Canada and abroad. In 1988 it merged with the Conservatory of Cinematographic Art and formed Cinémathèque Canada. The Institute maintains the CFI Film Library; conducts research; publishes books, monographs and resource material; organizes exhibits and film programs; and holds the biannual Ottawa International Animation Festival.

CANARA   Camara Nacional de Radio (Costa Rica).

candela (cd)   Also called **candle** or **foot candle.** Unit of intensity of light, expressing brightness in candelas per square meter. It replaced English foot candles. See also *lux.*

candid   Candid shot; unrehearsed, not posed, often unexpected, photographic shot.

canned music   Recorded, readily available music.

Cannes Film Festival   See *Festival International du Film-Cannes.*

CanP   Canadian Press.

cans   Colloquial term for earphones. See *headset.*

canted shot   A tilted photographic shot.

cap   See *lens cap.*

capstan   Revolving shaft (spindle) that pulls (and measures) the tape across the heads in a recorder/player at a constant speed. See also *pressure roller.*

caption   (1) Wording, lettering and all graphic material that appears on a television show. (2) See *subtitle.*

CAR   ITU country code for Caroline Islands (Palau).

CARA   Classification and Rating Administration.

CARACOL   Primera Cadena Radial Colombiana (Colombia).

carbon arc   Arc lamp using carbon arc electrodes for high power location/studio illumination; in projectors it is used for large cinema screens. See also *arc.*

cardioid   Cardioid microphone; a variable density microphone with a heart-shaped selectivity pick-up pattern.

carrier current   Low-power AM radio signals in non-broadcast use fed to power lines, pipes, and metallic networks to facilitate short distance radiation (in-house propagation).

carrier wave   Continuous electromagnetic wave emitted by radio or television transmitter. It may be modulated to carry data, music, voice, or picture signals. Receivers are tuned to receive only the carrier with a fixed wavelength for a particular station. See also *modulation*.

cartoon   An animated short film.

cartridge   Square-shaped container with rounded-off edges and feed and take-up reels. Audio- or videotape or motion picture film is inserted inside to facilitate loading, tape recording, replay, or projection.

cartridge projector   An automatic projector that accepts a cartridge-loaded film loop.

cassette   A container, similar to the cartridge, with two (feed and take-up) reels for holding magnetic sound and/or videotapes or films.

cassette recorder   A recorder that accepts and records on loaded cassettes.

cast   Dramatic personae; all characters, actors, performers, and extras who appear in a show or film.

casting   The selection and auditioning of actors for a specific part prior to a production. See also *audition*.

casting director   The person in charge of selecting actors (players) at a large studio or production house.

catchlights   Light reflections in the eyes of performers seen during tight close-ups.

cathode ray oscilloscope (CRO)   See *oscilloscope*.

cathode ray tube (CRT)   The principal tube in oscilloscopes, radar, and television receivers that changes electrical energy into light.

cathode ray tube projector   A special tube, operated at a high voltage, that produces a bright picture and is projected through a compatible optical system, used for large-screen projection. See also *Eidophor*.

CATV   Central aerial television; community antenna television.

catwalk   Elevated platform for handling lights on the set.

CAVIR   Camara Venezolana de la Industria de Radiodifusion (Venezuela).

CB   Citizens band.

CBA   Commonwealth Broadcasting Association (GB).

C-band   Frequency band of 3,700–4,200 MHz (USA, Russia).

CBC   (1) Canadian Broadcasting Corporation; Société Radio Canada-SRC. (2) Caribbean Broadcasting Corporation. (3) Commonwealth Broadcasting Conference (GB). (4) Cyprus Broadcasting Corporation.

CBG   ITU country code for Cambodia.

CBN   Christian Broadcasting Network (USA).

CBS   (1) Columbia Broadcasting System (USA). (2) Christian Broadcasting Service.

CBU   Caribbean Broadcasting Union (Barbados).

CC (closed-captioned)   (1) Television symbol indicating the program is subtitled for the hard-of-hearing or deaf; needs a special decoder. (2) Country code. (3) Closed circuit.

CCD   Charge-coupled device.

CC filter   Color compensation (correction) filter.

CCIR   Comité Consultatif Internationale des Radio Communications; International Radio Consultative Committee-IRCC (Switzerland).

CCITT   Comité Consultatif Internationale de Télégraphe et Téléphone; International Consultative Committee for Telegraph and Telephone-ICCTT (Switzerland).

C-clamp   A C-shaped metal clamp for hanging studio lamps on a scaffold.

CCR   Camcorder; portable video camera cassette recorder.

CCTV   Closed-circuit television.

CCU   Camera control unit. See *console*/1.

cd   Candela.

CD   Compact disc.

CD-A   Compact disc audio recording.

CDC&V   Check, double-check & verify. The principal "must-do," especially in news and documentation, even under the utmost pressure, fast time schedule, and deadline.

CdS meter Light meter with cadmium disulfide.

CEEFAX A British Broadcasting Corporation term for a dial-a-page broadcast teletext "see facts" service of news, travel information, weather maps, sports results, market reports, and cooking recipes for home television receivers. It uses two lines of the 625-line signal. See also *Viewdata*.

cel See *acetate*/1.

cell Cellular radio zone.

cellular radio Small scale radio telephone communication system, regulated by the Federal Communications Commission. For cellular radio telephone purposes, each county, city, or municipality is divided into zones (cells) of approximately 10-mile (16 km) radius. Each cell is assigned its own radio frequency (RF) in order to avoid signal interference.

celluloid See *acetate*/2.

cellulose acetate See *acetate*.

cement A liquid cellulose solvent used to join (glue) splices of film.

center focus antenna Also called **prime focus antenna.** A perfectly round antenna that has either a shallow or deep parabola.

Central Aerial Television (CATV) A system by which television signals are carried to home receivers by a coaxial cable that is hooked to a central antenna. Reception is usually much better and special programs may also be transmitted through the cable. There is a monthly fee for the service.

central control See *master control*.

CFI Canadian Film Institute.

CG (1) Character generator. (2) Computer graphics.

CGI Computer-generated image(s).

C.G.S. system Centimeter-gram-second system.

CH (Ch) Channel.

chain break Station break, station I.D.

change over cue See *cue*/2.

changing bag A lined, lightproof bag made of dark material with two elastic fitting sleeves that allows the camera operator or his/her assistant to load/unload undeveloped film into and from the camera magazine. Changing bags substitute for a darkroom on location.

channel (CH)   (1) A band of frequencies assigned and authorized for transmission. In standard broadcasting the channel is 10 kHz wide—5 kHz on either side of the carrier wave. In NTSC television the channel is 6 MHz wide; whereas cable television, on coaxial cable, carries a wide band of frequencies, to 400 MHz. (2) Electrical circuit for pick-up and transmission of sound.

channel width   See *bandwidth*.

character generator (CG)   Used in video-television graphics to create or generate letters and titles on a keyboard to be inserted electronically into the visual image. The letters may be highlighted, moved, manipulated, changed, stored and then retrieved.

charge-coupled device (CCD)   Image chip (80 times more efficient than photographic film), used in solid state video cameras to replace vacuum image tubes.

cheat   (1) An acting technique in television, when a shot is taken from a new angle, involving, for example, a cut from camera 1 to camera 2. The actor shifts his body to the same position in relation to camera 2 as it was in relation to camera 1. The audience, therefore, does not notice the change in angle. (2) Filmed presentation of successive or indicative elements of a movement that does not follow the entire movement in its true continuity.

checkerboard   (1) See *A-B rolling*. (2) Checkerboarding; in on-line videotape editing, a group of edits from one source are recorded prior to putting up another in a precise order following the edit decision list. See also *edit decision list—EDL*.

chest mike   See *lavalier*.

Chicago International Film Festival   Annual film festival honoring feature films, documentaries, shorts, animation, educational and student films and videos, presenting the Hugo Awards, Plaques and Certificates (USA).

china marker   See *marking pencil*.

CHL   ITU country code for Chile.

CHN   ITU country code for China.

choreographer   An artist and dancer who designs, and often conceives and arranges, movements and dance for a program or film.

CHR   ITU country code for Christmas Islands.

chroma A defined color saturation, its degree in the visible spectrum. In NTSC, chroma is sent as one signal, piggybacked upon the B&W monochrome video. Commonly known frequency is 3.58 MHz (3.57945 MHz actual). See also *hue*.

chroma key Electronic matting process in television, achieved through the use of the blue signal of the color camera.

chromatic aberration A lens property that causes passing light to reflect, as in a prism. The entering light is split up in the lens into its three components—red, green and blue rays—and comes into focus in different points beyond the lens (the red farthest away, the blue closest to the lens). The images formed do not coincide in size or in position. The resulting image on film will have color fringes.

chrominance channel Channel in the color transmission that contains chrominance information, derived from the original red, green and blue signals.

CID lamp Compact iodide daylight lamp—5500K.

CIE Commission International d'Eclairage.

Cindy Yearly award competition sponsored by the Association of Visual Communicators—AVC (USA).

CINE Council on International Nontheatrical Events (USA).

cine The French word for motion picture (cinema), popularly used in various combinations.

cinema Cinema house; movie (motion picture) theater.

Cinemacolor An obsolete color film process that first used a two-color, later a three-color system.

Cinema du Réel Festival International de Film Ethnographique et Sociologique; International Film Festival of Visual Anthropology and Social Anthropology for full-length and short films (France).

Cinemascope A wide-screen process, used extensively, which involves an anamorphic lens producing an image on 35mm film laterally compressed in a ratio of 2.35:1 (USA). See also *Panavision*.

Cinématèque Canada See *Canadian Film Institute*.

cinématographe A small, portable combined camera, printer, and projector unit designed in 1895 by the Lumière brothers, Louis (1864–1948) and August (1862–1954) of France.

cinematographer Cameraman/woman; director of photography.

cinema van   Specially equipped van containing either film or film projection equipment. See also *camera car, camera truck, remote van* and *shotmaker.*

cinéma-vérité   Also called **direct cinema.** A way of filming real-life scenes without the use of elaborate equipment or a large crew, i.e. using the actual setting, existing lights, handheld camera, and portable sound equipment. Applies also to video.

Cinerama   Originally a wide-screen process using three 35mm cameras and three projectors on a wide screen. The process now employs a single 70mm camera and one projector on a wide screen in a ratio of 2.2:1 (USA).

CIP   Centre d'Information de Presse (Belgium).

circle in, circle out   Iris in and iris out. See *iris wipe.*

CKH   ITU country code for Cook Islands.

clap board   Clap stick; a pair of short hinged boards clapped together to provide a sharp cue point visible on the picture and a corresponding audible cue on the sound track. It helps to establish synchronization in the editing process. Most modern film cameras are manufactured with a built-in cueing system. See also *slate* and *beep.*

clapper   (1) Clap board/sticks. (2) A member of the camera crew who operates the clap board at each take.

clapper sticks   See *clap board.*

Clarke belt   Arthur C. Clarke belt; geostationary or geosynchronous orbit. See *synchronous orbit.*

class A time   See *peak time.*

claw   A device in the camera and small projector that engages the sprocket holes in the film and pulls it down in an intermittent motion.

cleaning   (1) Film cleaning; an operation to remove dirt, dust, fingerprints, and other impurities from film surfaces. Done either in small manual or high-speed motorized cleaning machines by rolling the film through pressurized pads soaked with cleaning solvent. Cleaning and lubrication also reduces film noise and minimizes damage. Modern methods use ultrasonic cleaning. See also *lubrication.* (2) Equipment cleaning; essential maintenance of hardware, including gates, lenses, rollers, magnetic heads, and various moving parts in cameras, recorders, projectors, and other equipment.

clearance   Permission to use copyrighted material.

clear channel   Broadcast channel on which an AM broadcast station transmits over a wide area, and which is cleared of relative interference. Clear channel usually means no other station is on the same frequency.

client   Individual, firm, company or agency advertising in radio, television or cinema houses; a sponsor.

Clio Award   Annual best video award held in New York City for television/cable productions (USA).

clip   (1) The sudden cut-off of the audio part of a program. (2) A short tape or film section used as an insert. (3) A piece of film clipped off in the editing room. See also *cut*/5 or *trim*/2.

clipstrip   Broad, even lighting, using quartz lights with barn doors, placed on overhead bars or on stands for lighting cyclorama or background scenes.

CLM   ITU country code for Colombia.

CLN   ITU country code for Sri Lanka.

clone   Term for an imitated, successful broadcast program.

closed circuit (CC)   Television program transmitted to a specific (selected) audience usually by cable or directional radiation. It is not telecast to the general public.

closed set   Sound and/or videotape or film production closed to the public and observers, accessible only to crew and staff.

close shot   Close-up.

close-up (CU)   A shot usually taken by a narrow-angle (telephoto) lens, where the subject fills the frame and appears in very close range. See also *gros plan*.

cm   Centimeter; 1/100 of a meter, 1/10 of a decimeter, or 10mm; equals approximately 3/8 of an inch.

CM   Circuit multiplication.

CME   (1) ITU country code for Cameroun. (2) Center for Media Education (USA).

C-mount   The accepted screw threading used in small gauge, 16mm film and many small video cameras to hold and/or interchange lenses. See also *bayonet mount*.

cm/s   Centimeter per second; equals approximately 3/8 inch per second.

CMTV   Country Music Television cable network (USA).

CMX (CMX 6000)   Trade name of a magnetic storage system for video-tape editing. Trademark of the CMX Corporation.

CMY   Cyan, magenta, yellow. See *subtractive color process.*

CNA   Central News Agency (Republic of China).

CNBC   Consumer News & Business Channel cable television network (USA).

CNIRTM   Camara Nacional de la Industria de la Radio y Television de Mexico.

CNN   Cable News Network (USA).

CNNI   Cable News Network International.

CNR   (1) ITU country code for the Canary Islands. (2) Control Nacional de Radio (Costa Rica).

coating   (1) Lens coating with a thin, transparent material that reduces light reflection and flare. Coating also improves light transmission through the lens. (NEVER use ammonia-based cleaning agents on lenses—it removes the coating.) (2) The light-sensitive emulsion or the magnetic material on the tape or film base. (3) See *bloom/2.*

coaxial cable   Specially designed cable used to carry picture or high-frequency signals (audio, video or data) consisting of two concentric cables in a housing shielded by an insulating medium.

Codec   Colloquial for digital coder and *decoder.*

code numbers   Matching (identical) numbers printed on the edge of the picture and sound track in the editing process to facilitate perfect synchronization.

coding machine   An apparatus that prints code numbers on film.

COG   ITU country code for the Congo.

cold   A program that begins without an introduction, announcement, or theme.

cold light   Fluorescent light; not suitable for color film production.

cold reading   Announcement read without previous rehearsal.

color   (1) Television or film recording, transmission, and projection of color pictures. (2) An artistic environment, a setting to create atmosphere (color) in a program or show.

color balance   True reproduction of color in television or film, adjusted to a white reference.

color bar   Color test bar; color chart or an electronically generated six-color part of the color test signal consisting of vertical bars of red, green and blue (primary colors) and yellow, magenta, cyan, black, and white. Also has special NTSC test signal for color encoder testing (USA).

color burst   A short color burst or pulse of the subcarrier frequency to trigger each line in the receiver to reproduce the exact color. Present only in the *NTSC* and *PAL* systems, but not in *SECAM*.

color contrast viewing filter   A specific filter used before filming in color to determine highlights and shadow detail. See also *panchromatic viewing filter.*

color conversion   See *daylight conversion.*

color correction   (1) Videotape color correction during electronic editing. (2) Correction and color matching in the film camera using filters. (3) Laboratory correction and matching of colors of exposed, developed film.

color dupe negative   Color duplicate negative; color negative film made from an original negative, used for release printing.

color film   Non-monochrome

color filter   See *gelatin filter.*

color internegative   Color negative film made from a color positive original, used for making release prints.

colorization   Computerized process that adds color to black and white (B&W) films.

colorized film   Old, sometimes classic, B&W films that have been "modernized" through colorization.

color master positive   A positive color print made from a color negative original.

color negative   Color camera film with negative images.

Colorplexer   Encoder; device that transforms the original color signals and related information into the necessary coded form for transmission (*NTSC* or *PAL*).

color print   Positive color print film made from the original or from a color internegative.

color reversal film   Color film with positive images.

color reversal intermediate (CRI)   Color internegative film made from the original negative by reversal process.

color saturation   See *saturation*.

color sensitivity   Sensitivity of the film to a certain portion of the spectrum.

color television   Non-monochrome television system using the three primary colors—red, blue, and green. Various methods are used to produce these colors. Three systems have been developed as standards and are used world-wide: NTSC, devised in the United States of America; PAL, devised in Germany; SECAM, devised in France. NTSC became the standard color television system in the U.S. in 1953. See also *NTSC, PAL,* and *SECAM*.

color temperature   A measurement of a light source's color value, expressed on the Kelvin scale. Refers to the relative redness and blueness present. Color television lights can range between 2800 and 3400 K. Most color film requirements are at 3200 K. See *Kelvin scale*. See also *daylight*.

color temperature meter   A device, similar to the exposure meter, that measures color temperature on the Kelvin scale, facilitating the matching of the color temperature of lights with the production system.

COM   (1) ITU country code for Comoros. (2) Communication Division of UNESCO. See *United Nations Educational, Scientific and Cultural Organization*.

combination/combo shot   A photographic shot with one character close-up in the foreground, and another behind him in a medium or long shot.

comedy/variety program   A broadcast program consisting of humorous scenes, situations, monologues, one-liners and various singers and dancers.

comet tail   Also called **image retention.** A bright image moving in the scene, leaving a path, like a comet's tail streaking and trailing—an imaging defect in older Plumbicon video camera tubes.

COMFER   Comite Federal de Radiodifusion (Argentina).

commag   Combined magnetic-track (sound) film.

commentary   A running treatise of comments during an event, show, or program; narration.

commentator   The person who announces and describes the events being broadcast. See also *narration*.

commercial   Colloquial term for advertisement broadcast or screened. They may be live, taped, filmed, animated, or any combination of the above.

communication arts   The art of mass communication; an academic term comprising the art and knowledge of the press, radio, television, and motion pictures.

communication satellite   See *satellite communication*.

Communication Satellite Corporation (COMSAT)   Formed in 1962, the organization provides digital and analog voice, data, and video communications services via satellites to international communications carriers and television networks worldwide. (USA). See also *Intelsat*.

communications payload   The transmitting and receiving payload carried by a satellite.

Comopt   Combined optical-track (sound) film.

compact disc (CD)   A compact-size recording disc employing digital signal processing and laser beam technology to transmit audio, video, and data signals.

compatible color   Also called **compatibility.** Color television transmission that has sufficient brightness control to be received on monochrome (B&W) sets.

compensator   See *voltage regulator*.

compère   A broadcast personality who organizes, announces, and introduces a program (usually entertainment) and the performance. See *DJ* or *MC*.

complete video   See *composite video*.

composite print   Composite master; a combined fine grain positive film print containing both the synchronized picture and sound track. See also *married print*.

composite video   Also called **complete video** or **VBS.** Video blanking signal complete with chrominance, luminance and sync information.

composition   See *picture composition*.

compression   Two or more streams of information squeezed (compressed) into one channel. See also *digital compression*.

COMPTU   Council of Motion Picture and Television Unions (USA).

computer generated image(s) (CGI)   Images other than drawings, graphs, cutouts, or photographs created (generated) and then arranged and manipulated by a computer.

COMSAT   Communications Satellite Corporation (USA).

CONATEL   Conseil National des Télécommunications (Haiti).

condenser mike   A sensitive microphone with wide dynamic range, excellent frequency response, and no distortion. It is unobtrusive, small, and easy to handle.

conforming   See *negative assembly*.

console   (1) A centralized video control panel for a camera control unit (CCU) and the control of sound and visual production in a multi-camera set up. It is equipped with sound and video monitors, electronic devices for level and signal control, echo and special effects. An intercom system connects it with studio personnel. (2) A control panel for sound recording and mixing in a film studio.

constant speed motor   See *governor motor*.

consumer grade   Video camera grouping indicating cameras for amateur (non-professional) use by the general public. See also *broadcast quality* and *industrial grade*.

contact printing   A fast printing method in which negative and positive films are printed together with emulsion sides facing each other, producing the same image size as the original. See also *optical printer* and *step printer*.

container   See *film can*.

continuity   (1) Even, logical flow of program content or film. (2) All the material presented on the air, including short announcements and commercials, and apologies. See also *script*.

continuity cutting   Film editing in the conventional style with emphasis on the continuous flow of action. See also *dynamic cutting*.

continuity department   Broadcast station department where continuity material is written before production approval.

continuity girl   See *script supervisor*.

continuity studio   See *announcing booth*.

continuity writer   The person who writes and prepares part or the entire program and organizes it in continuity.

continuous loop   See *film loop*.

continuous projector   Projector employing a continuously running film loop; needs no rewinding.

continuous shot   Uninterrupted camera shot following the action. See also *follow shot*.

contour   See *loudness control.*

contrast   The ratio of two brightness levels; the difference between black and white; the darkest or lightest part of a television picture or film.

contrast filter   Lens filter used to change relative brightness values. See also *correction filter* and *haze filter.*

controller of programmes   Program manager; director in charge of the largest department of a broadcasting house who plans and executes programs. He/she also supervises the quality and overall balance of productions. British term.

control panel   See *console.*

control room   The room, usually overlooking the broadcast studio, from which a program is controlled, coordinated, and directed. It accommodates the director, technical director (switcher or vision mixer), audio and video engineers and assistants.

control strip   A filmstrip exposed to stepped density scale in a strictly controlled fashion.

control track (CTL track)   Supplementary on videotape, a track for 30 Hz to 60 Hz speed synchronization signals.

conversation   See *dialogue.*

conversion filter   See *daylight conversion/2.*

converter   See *standards conversion;* also *line rate converter, line store converter,* or *transformer.*

cookie (kukie)   See *cucalorus.*

coop   Colloquial for *mercury vapor lamp.*

COPER   Agencia Noticiosa Corporacion de Periodistas (Chile).

copy   All material prepared for a program to be read on the air.

copy editor   Editor who corrects, edits, and revises copy for broadcast.

copyright   Branch of law covering ownership and control of literary material, photographs, motion pictures, musical compositions, videotapes, video programs, or other creative works registered with the U.S. Copyright Office (Library of Congress), or with the appropriate office elsewhere, signatories to the Universal Copyright Convention (Bern, Switzerland, 1886, and Geneva, Switzerland, 1952). The statutory copyright period is for the life of the author plus 50 years for works created on or after January 1, 1978. At the expiration of the copyright the work becomes public domain.

Copyright Royalty Tribunal (CRT) A three-member review board which, from time to time, reviews and adjusts compulsory royalty rates. The members of the Tribunal are appointed by the President and confirmed by the Senate (USA).

core A hollow central part molded of plastic, metal, or wood, on which videotape or film is (rolled) wound.

Cork International Film Festival Competitive international film festival for feature and documentary films held in Cork, Ireland.

corporate video In-house video programs and department of a corporation, used mainly for promotions, public and consumer relations, product advertisements, training, and internal security.

correction filter Lens filter that changes the color quality of light, allowing B&W film to record all colors in relative brightness. See also *contrast filter* and *haze filter.*

correct speed Sound and picture running in perfect synchronization.

correspondent Broadcast (news) personnel stationed away from headquarters, usually in foreign capitals and/or at centers of international organizations, such as the UN and World Health Organization, who send regular news reports and commentaries to the home office.

costume Also called **wardrobe.** A complete set of garments, including jewelry and accessories, worn by actors to accentuate period or historical plays and programs. See also *hand props* and *props.*

costume designer Designer who plans and develops the costumes (by drawing sketch designs) for a show or film. He/she also selects fabrics and supervises the assembly and fitting of garments.

coulomb The unit of electric charge; the quantity of electricity transferred in one second per one ampere. Named after Charles A. de Coulomb (1736–1806), French physicist.

Council of International Nontheatrical Events CINE; a Washington, DC-based organization founded in 1957 that introduces and now regularly distributes and sends specially selected films and video programs to specific international events abroad. (USA) See also *Golden Eagle.*

countdown Videotape or film countdown, as the Academy numbers flash on the preview monitor in one second intervals for precise cueing.

counter programming Broadcast scheduled against a competitive program at the same time.

coverage   (1) The area and the number of radio and/or television homes a broadcast station's signal is able to reach. (2) The recording and filming of actual (news) events. (3) The filming of various angles with more than one camera. See *multicam*.

cover shot   A large area picture, photographed usually with a wide-angle lens, that identifies a location and gives basic orientation for a program or event. See also *establishing shot, opening shot,* and *orientation shot*.

CPB   Corporation for Public Broadcasting (USA).

CP filter   Color printing filter.

CPM   Cost per thousand.

cps (c/s)   Cycles per second.

CPV   ITU country code for Cape Verde.

crabbing   See *trucking*.

crab dolly   A wheeled camera mount with steering controls that enable it to be moved in different directions even in confined areas.

cradle head   A cradle-shaped, sometimes geared, tripod head.

crane   A specially built vehicular camera boom that can be moved in all directions and lowered or raised high above a scene.

crank handle   See *hand crank*.

crawl   Also called **drum** or **roller title.** Titles and credits of a television or film program mounted on a drum and rolled to move (crawl) slowly up on the screen. Old-fashioned, now replaced by electronic computer lettering.

credits   List of names and functions of production personnel and performers who participated in the planning, production and execution of a program, appearing at the beginning and at the end of a program or film. Radio program credits are voiced over the air.

creeping   Slow, irregular movement of the picture.

creeping title   See *crawl*.

crew   The production team; the technical personnel. See also *unit*.

crew call   Call time for the production personnel. See also *call*.

CRI   Color reversal intermediate film.

crib card   See *shot sheet*.

CRO   Cathode ray oscilloscope.

cropping Reduction of an image by the use of masks or a smaller aperture.

cross fade The gradual dimming (fade out) of a picture and/or sound, and the inclusion of another picture and/or sound over the previous one. See also *dissolve* and *segue*.

cross hairs Popular name of a cross representing the exact center of the film frame etched into the ground glass of the camera viewer.

cross light Also called **kicker.** Illumination coming from the side, across the scene.

crosstalk Leakage of audio, video or RF signals between two channels, caused by the proximity of receiving devices, poor shielding, or partial component failure.

crowfoot See *triangle*.

CRT (1) Cathode ray tube. (2) Copyright Royalty Tribunal.

CRV Component Video(disc) Recording.

crystal See *quartz*.

crystal black See *black/4*.

crystal control Also called **crystal sync.** (1) A crystal (slice of quartz) vibrating at a reliable constant frequency, used to assure stable scanning. (2) In motion picture sound filming it maintains a steady, perfect synchronization between the sound and picture without the use of cable.

Crystal Radio Award See *National Association of Broadcasters*.

CS Cinemascope standard. See also *BH* and *KS*.

c/s (cps) Cycles per second.

CSA Casting Society of America.

CSC Canadian Society of Cinematographers.

CSI lamp Compact sourc iodide lamp 4000K.

CS perforation 35mm Cinemascope perforation.

C-SPAN (CSP) Cable Satellite Public Affairs Network (USA).

CST Central Standard Time (USA).

CT Computer technology.

CTAM Cable Television Administration and Marketing Society (USA).

CTI ITU country code for Côte d'Ivoire.

CTK   Česka Tisková Kancelař (Czech Republic).

CTL track   See *control track.*

CTR   ITU country code for Costa Rica.

CTT   Empresa Publica dos Correios e Telecomunicações (Cape Verde).

CTV   (1) Česka televise (2) Community television.

CTW   Children's Television Workshop (USA).

CU   Close-up.

CUB   ITU country code for Cuba.

cubicle   Announcing booth.

cucalorus   Also called **kukaloris** or **kukie.** An irregular shadow pattern (cut-out) placed in front of a spot light and projected on a dull, flat surface.

cue   (1) Any sign or signal indicating the start of speech and/or action on the set in a studio or on location. Cues may be voiced or hand signalled if sound is restricted. (For "Q" from the Latin *quando*— when.) (2) The line up of records or transcriptions at the required band or cut that may be played without delay. (3) Broadcast station or network identification on the hour or at the close of a program.

cue card   Also called **idiot card.** A large card or sheet with highly visible view lines or script held near the camera to aid the announcer or actor.

cue light   British term for **tally light.**

cue marks   Marks that indicate the approach of the end of the videotape or film. In television it is electronically effected; in film it is printed in the laboratory.

cue sheet   A list of specific cues tabulated in editing on a sheet for a given show or film. See also *cue card.*

CUFC   Consortium of University Film Centers (USA).

cushion   Additional material prepared for use in case the broadcast program runs short.

cut   (1) A cue, a signal to stop all speech and/or action. (2) A special part, a selection on a record (disc) or transcription. Also called **band.** (3) A command to switch from one television camera to another, i.e. "Cut to two". (4) An instantaneous transition in a film from one scene to another by joining (splicing) two shots together. (5) A short section of a film clipped off or removed in editing—an out take.

CUT Coordinated Universal Time/Universal Time Coordinate-UTC.

cut-away A fast shot, usually a neutral shot away from the main action. A reaction shot, a crowd or street scene, used mostly as a transition.

cut-back A short return to a previous scene after a brief cut-away.

cut in Special message or news flash inserted into an ongoing broadcast program. See also *insert.*

cutter Film editor.

cutting The editing of the film and/or tape.

cutting copy See *workprint.*

cutting room Editing room.

cut-out (1) Small figures cut out of paper, cardboard, wood, or other material and used in animation. They are manipulated and photographed in such a way that an illusion of continuous motion is created on the screen. (2) A two-dimensional shape of a person, building, etc. cut to scale.

"Cut your darlings" Expression in the film editing room after some memorable and beautiful scenes or shots are edited out, ending up on the editing floor or in the bin, usually because they would throw off the rhythm of the film and alter the flow of the scripted order.

CV Comedy/variety program.

CVA ITU country code for the Vatican City State.

CVN Cable Value Network (USA).

CWA Communication Workers of America.

CXR Carrier.

cyan Red-absorbing, blue-green (minus red), subtractive primary color, used in three-color process. See also *subtractive color process.*

cyc Abbreviation for cyclorama.

cycle See *hertz.*

cyclorama (cyc) A large continuous piece of cloth curved around the scenery to form a backdrop or to indicate sky.

cyc strip   High-intensity strip lighting, using quartz lamps, placed on the floor or hung overhead to give even illumination to a cyclorama.

CYM   ITU country code for Cayman Islands.

CYP   ITU country code for Cyprus.

CZE   ITU country code for the Czech Republic.

D

D (1) ITU country code for Germany. (2) Day; daytime.

D/A Digital to analog. See also *A/D*.

DAB Digital audio broadcasting.

daguerreotype Direct photography on a copper plate with a silver iodide surface. Invented by painter Louis J. M. Daguerre (1789–1851) in France. The development of the latent image was achieved with a mercury vapor process.

dailies Also called **rushes.** First laboratory prints of the exposed film after a day's shooting. These are one-light review prints, not necessarily in script continuity, but in order of scenes as they have been filmed.

darkroom A light-controlled room, similar to one used in photography, usually equipped with a safety light, where film is rolled, loaded, and reloaded. See also *changing bag*.

DAT Digital audio tape.

DATE Digital Audio for Television.

day-for-night (D/N) See *night effect*.

daylight Sunlight; midday sunlight is equivalent to 5,400 Kelvin (average). See also *tungsten*.

daylight conversion (1) Dichroic filter (mirror) that converts incandescent or quartz lights into daylight. Used extensively in color television. (2) Photometric filter used in color film production for changing the color temperature of the light source to match it with the film used. See *tungsten*. See also *blue/2* and *orange filter*.

daylight loading spool Metal film spool with solid flanges to protect tightly wound, unexposed film from light. This spool may be loaded carefully in regular (subdued) lighting conditions without using a darkroom or a changing bag.

daytime serial (DS)   A weekly television program, 30 or 60 minutes long, telecast during the day. Usually a serial drama. See also *soap operas*.

dB   Decibel.

DB   Delayed broadcast.

DBS   Direct Broadcast Satellite (direct satellite broadcast).

DC   Direct (unidirectional) electrical current. See also *AC*.

DCME   Digital Circuit Multiplication Equipment.

DDP   Deutscher Depeschen-Dienst (Germany).

dead   (1) Equipment that is not functioning, either because it is faulty or because it has not been turned on; i.e. "dead mike". See also *hot camera, hot mike*. (2) Studio or set or a place with little or no reverberation.

dead spot   (1) Dead air; unintentional pause (silence) on the air. (2) Flat, actionless space on the set that upsets composition.

decibel (dB)   One-tenth of a bel. A logarithmic unit expressing power ratios, often used to measure relative intensity of sound or other electronic signals.

decimeter   10 centimeters, or one-tenth of a meter. Equals appr. 3.9 inches.

decoder   A color television receiver device that separates the encoded signal (NTSC, PAL, or SECAM) into the original red, green and blue (RGB) signals.

decor   Decoration, set, scenery.

definition   Degree of reproduction detail of an image; the sharpness of television picture or film. See also *resolution*.

de-focus   To bring subjects out of focus. See *focusing*.

degausseur   Powerful eraser for large reel magnetic tape and film.

delayed broadcast (DB)   Broadcast program videotaped and aired later than its original (network) schedule.

demagnetizer   See *eraser*.

demonstration (demo tape, demo film)   Demonstration recording or film mostly used for planning and preparing broadcast commercials.

density   Photographic density; the factor indicating opacity, the extent to which light is absorbed by an object.

depth of field   The field in front of the camera in which the objects placed in different distances register in sharp focus. Depth of field depends on focusing distance and the focal length and aperture of the lens in a given lighting condition.

depth of focus   The range of the camera lens within which the image can be moved and still appear well defined, in sharp focus.

desk stand   See *table stand.*

deuce   A two-lamp light unit in a single housing.

developer   Chemical solution used in laboratory film processing.

development   Refers to (1) Script. (2) Character. (3) Program. (4) Film.

DG   Dramatists Guild (USA).

DGA   Directors Guild of America.

DGA Awards   Awards presented by the Directors Guild of America for outstanding achievement in television direction, commercial direction, and theatrical direction. See *Directors Guild of America.*

DGRTN   Direccion General de Radiodifusion y Television Nacional (Guatemala).

DH   Dubray-Howell.

diagonal cut   Direction of a splicer, thus a cut, which is diagonal rather than straight. Mostly used for cutting magnetic (sound) film. See *straight cut.*

dialogue   Conversation; conversational form of a script or scene, as opposed to narration or commentary.

dialogue director   The person in charge of a film's talking scenes. Common position in Europe.

dialogue track   Soundtrack containing dialogue portion. See also *M&E track.*

diaphragm   Also called **iris.** Adjustable lens opening device that controls the amount of light passing through the lens. F-stops are used to calibrate the openings.

dichroic mirror (dichroic filter)   Mirror-like filter that reflects a selected wavelength of the color spectrum while transmitting others, thus selecting colors. Used in color television systems.

diffuser   (1) A fiberglass, gauze, blotting paper, or other material placed in front of a studio lamp to soften light. See *butterfly.* (2) Gauze or similar material placed in front of a lens for soft or special effects, e.g. to simulate an antique painting.

digital audio, digital recording, digital sound, digital tape, digital video effect Recording and production technique employing digital technology, eliminating signal distortion or noise. Digital audio tape must be converted to analog in order to be used conventionally.

Digital Beta New version component video format providing direct digital recording. See also *Betacam.*

digital compression Method of compressing (squeezing) images for editing purposes and for transmission, then decompressing them to normal reception size.

Digital Satellite Radio (DSR) Broadcasting method that allows transmission of up to 16 digital radio stations on a single transponder. Used mostly in Europe.

digital television (DT) Based on digital technology, using the common language of binary code, it improves sound and picture quality by "cleaning up" incoming signals. Also permits storage and manipulation of pictures, zooming, freeze frame, and changes to other types of digital data.

dimmer Instrument that controls the amount of electric current flowing into lights, thus allowing for variations in the intensity of illumination.

DIN International standard of the emulsion speed of photographic and motion picture film, established by the Deutsches Institut für Normung (Germany). See also *ASA, ISO.*

dinky inky See *inky dinky.*

diode Two-element tube.

diopter The unit of power of a lens; the reciprocal of its focal length in meters. Positive converging lens is marked (+), negative diverging lens is marked (–).

diopter lens See *split field lens.*

direct broadcast satellite (DBS) Satellite with powerful Ku-Band that broadcasts only four or five channels directly to an end user.

direct cinema See *cinéma vérité.*

direct current (DC) Electrical current (unidirectional) that flows in the same direction. See also *alternating current.*

directional antenna An aerial with a limited (specific) angle of acceptance.

directional microphone Microphone with a pick-up pattern in a specific direction. See *bi-directional mike, omni-directional mike,* or *uni-directional mike.*

director (1) The person in charge of all production (creative) elements, composition, action, and coordination of the script for a radio or television show. (2) The creative coordinator of a film script in control of performers, action and production crew on the set or on location. The person contracted by and responsible to the producer.

director of photography (DP) Head of the camera crew in charge of photography, pictorial composition, and lighting of a film production. See *cameraman/woman, camera operator.* See also *lighting cameraman/woman.*

director's finder Director's viewfinder; optical device similar to a viewing lens used by the director for selecting angles, frames, and picture composition.

Directors Guild of America (DGA) A union organization that began in 1936 when a dozen motion picture directors in Hollywood formed the Screen Directors Guild, then admitted assistant directors a year later. In 1960 the Screen Directors Guild merged with the Radio and Television Directors Guild and was named the Directors Guild of America.

The Guild today represents motion picture directors, radio and television directors, associate directors, assistant directors, unit production managers, stage (floor) managers, and production assistants, and publishes *Action* magazine and the annual *Directory of Members.*

The Guild also organizes the annual DGA Awards presentations for television direction, commercial direction, and theatrical direction, and presents the DGA Lifetime Achievement Award. The Preston Sturges Award is given for outstanding achievement in both writing and direction.

direct print Color print made in one step, directly from the original film. See also *one-light print.*

direct recording See *single system.*

direct to disc Direct recording.

direct viewfinder See *viewfinder/2.*

direct wave Electromagnetic waves that travel from the transmitting antenna to the receiving antenna in a direct (line-of-sight) path. See also *ground wave* and *sky wave.*

DIS The Disney Channel cable television network (USA).

disc   (1) Gramophone record (disk). (2) Computer disc (diskette). See *compact disc* and *video disc*.

disc jockey (DJ)   A radio MC or compere.

discussion program   See *informational program*.

dish pan   Also called **dish.** Colloquial term for a dish-shaped parabolic reflector used for microwave (relay) transmission.

display tube   See *picture tube*.

dissolve   Also called **lap dissolve.** Gradual appearance (fade-in) of a picture as a previous one is being taken out (fade-out); the two overlap briefly during transmission. In television it is controlled electronically from the console; in film production it is achieved by a laboratory process. See also *mix/2*.

distance fog   Fog or haze appearing in long shot (LS) distances.

Distinguished Service Award   See *National Association of Broadcasters*.

distortion   (1) Unwanted deterioration or compression of a signal, either audio, video, RF or data. (2) An unintentional optical distortion caused by a lens defect. (3) Intentional distortion in picture composition in which a near object appears out of proportion, an effect achieved by the use of a wide angle lens.

distribution   The rental, lease, or outright sale of films and video programs. See also *syndication*.

ditty bag   A small, open-top canvas bag attached to tripod legs to hold useful accessories like camera tape, scissors, etc.

DJ   Popular abbreviation for *disc jockey*. Used also as veejay. See also *VJ*.

DJI   ITU country code for Djibouti.

DLF   Deutschlandfunk (Germany).

DLP   Demokraattinen Lehdistopalvelu (Finland).

DMA   ITU country code for Dominica.

DMA Rating   Designated Market Area rating. See also *ARB*.

DMR   Digital mobile radio.

DN   Documentary news program.

D/N   Day for night.

DNC   Direccion Nacional de Comunicaciones (Uruguay).

DNK ITU country code for Denmark.

docking Dockable camera; portable video camera designed to dock directly with specified professional recorders, also attaches to professional and broadcast Betacam recorders with an adapter. Used mainly in one-man/woman EFP and ENG applications to ensure maximum mobility.

docudrama Generic term for a dramatized documentary with fictionalized facts.

documentary (1) Documentary program or film. Realistic treatment of a subject in terms of actual events and background without the use of a fictionalized story. (2) A tape (sound or video) or film, usually of a short subject, depicting factual happenings in their true form, customarily accompanied by natural sound effects and a separately or simultaneously recorded commentary. See also *news program*.

Dolby system Trade name for a popular audio control system, designed and developed by Raymond Dolby, American inventor, to achieve better fidelity by reducing noise levels in audio recordings and reproductions.

dolly (1) A vehicular camera or microphone support that enables the camera or microphone to be wheeled in different directions. (2) Movement of the camera on a dolly toward the subject—"Dolly in," or away from the subject—"Dolly out" or "Dolly back." See also *trucking*.

dolly shot Also called **trucking shot.** Dollying; shot taken while the camera on a dolly is in motion. See also *crabbing*.

DOM ITU country code for the Dominican Republic.

DOMSAT Domestic satellite; colloquial for satellite signals relayed domestically within a country. See also *COMSAT* and *INTELSAT*.

dope sheet Film analysis prepared by the film library.

Doppler effect Doppler principle; as the distance between the wave motion (sound or light) source and the observer increases or decreases, the frequency of the wave increases or decreases respectively. Named after Christian Johann Doppler (1803–1853), Austrian mathematician and physicist.

double A talent with some special ability or likeness who takes the place of the principal actor during set up, in dangerous scenes, or in takes where a particular skill is needed. See also *stand-in* and *stunt man/woman*.

double exposure Two successive images superimposed on the screen or on the same piece of film. The effect on television may be achieved electronically; on film it may be done directly in the camera or by a laboratory process. See also *superimposition*.

double perforation Also called **double perf.** Motion picture film stock that has been perforated (sprocket holes) on both edges in the manufacturing process. See also *single perforation*.

double system Double system sound recording; a method by which the picture and the synchronous sound are filmed and recorded separately (not on the same film stock) to be combined later in editing and printing. See also *single system*.

down and under See *background/1*.

down converter Also called **front end.** Device that transfers high satellite frequencies to lower frequencies suitable for regular television receivers.

down-link Orbiting satellite signal path to the television receive-only antenna (TRVO). See also *up-link*.

downstage The part of the studio set closer to the camera(s).

downtime Equipment service, repair, or maintenance time that halts a production.

dowser An electrical device used during film projection that cuts off the beam of light to facilitate changeover from one projector to another.

DP (1) Director of photography. (2) Dramatique personae. See *cast*.

DPA Deutsche Presse Agentur (Germany).

dramatique personae See *cast*.

dramaturge A dramatist or writer who revises, arranges, and often selects dramatic material for production. See also *story editor*.

dress (1) Costume worn by the performers. (2) To dress the set, furnish the scene.

dressing light See *background light*.

dress rehearsal Final rehearsal with costumes, cameras, sets, lighting, and props. See also *camera rehearsal*.

drive-in theater Open air movie theater (in declining existence) where the audience views film programs from their parked cars and the sound is supplied by individual small speakers to each vehicle or by low-power radio transmissions picked up by the radio in the automobile.

drop A large piece of background material made of colored or painted canvas; a backdrop.

drop out Loss of a sound or video portion on the recording caused by an imperfection or impurity (dust); also caused by the tape's oxide flaking or being scratched off. See also *bloop.*

drop shadow Also called **off-set title.** Shadow "dropped" along the letters in titles so they will stand out.

drum (1) A loosely rotating metal cylinder on a spring or flywheel on which sound tape or film is wrapped to ensure steady tension. (2) See *crawl.*

dry carrel Study carrel with no audiovisual hardware. See also *wet carrel.*

dry run Rehearsal in the studio without the involvement of the program's technical aspects and with minimal possible interruptions.

DS Daytime serial.

DSR Digital satellite radio.

DSS Direct satellite system.

DT Digital television.

DTB Digital television broadcasting.

DTH Direct-to-home satellite broadcasting. See *direct broadcast satellite (DBS).*

dual track See *half track.*

dual track recorder Two-track recorder; a monophonic recorder that records and/or plays half of a standard ¼" tape in one direction, and the other half in the opposite direction.

dub Videotape copy (for television use). See also *dupe.*

dubbing (1) The transfer of the sound from one recording to another; from disc to tape, or vice versa. (2) The mixing of several sound tracks and recordings on a single track. (3) Lip synchronization of a performer's voice in the sound studio with the separately or previously shot picture. See *post-synchronization.* (4) The addition and recording of sound to the visual image of tape or film. (5) The substitution of foreign language dialogue in the original program or film as opposed to subtitles. Due to small size of television receivers, dubbing is often used during telecasts of foreign films or programs.

dull Uninteresting, unimaginative script, performance, and/or picture composition.

dulling spray Pressurized spray in a can used to eliminate or reduce glare and reflection of light on shiny surfaces.

dupe (1) A video or audio tape copy. See also *dub*. (2) Duplicate negative; negative motion picture film made from a positive original.

duplicate negative See *dupe/2*.

duPont, Alfred I. Columbia University Awards Annual presentation by the School of Journalism, Columbia University, recognizing outstanding work in news and public affairs. (USA).

DVE Digital video effects.

DVR Digital video recording.

DVTR Digital videotape recording.

DW Deutsche Welle, foreign service broadcast (Germany).

DX Long distance.

dyn See *dyne*.

Dyna lens Trade name for a camera lens designed to correct vibrations in moving shots (cars, helicopters, boats, etc.) without the use of specific mounts. See also *helicopter mount*.

dynamic cutting Film cutting (editing) that places contrasting sequences and shots together not in continuity, but in such a way that the dramatic effect is created in the viewer's mind. See also *continuity cutting*.

dynamic microphone (1) A pressure microphone that receives sound vibrations on a sensitive diaphragm. The electrical impulses are then transformed in a moving coil. (2) A rugged, small-size microphone used both in studio and remote (location) programs.

dynamic range The ratio between the loudest and softest sounds a recorder/playback machine is able to produce/reproduce without distortion.

dyne Dyn; unit in the C.G.S. system that equals the force that would give a free mass of one gram (g) an acceleration of one centimeter (cm) per second.

E

E ITU country code for Spain.

early bird See *Intelsat*.

earphones See *headset*.

earth station TRVO antenna.

Eastman color A widely used integral tri-pack negative/positive color film process for 35mm and 16mm film production. Made in the United States since 1952.

EAVE Les Entrepreneurs de l'Audiovisuel Européen; Media Program of the European Union.

EBR Electron-beam recorder.

EBS Emergency Broadcast System.

EBU European Broadcasting Union/Union Européenne de Radiodiffusion-UER.

E.C. (1) Electric current. (2) Electronic church. (3) Electronic cinema.

echo Reverberation of sound waves magnified and delayed to create a "hollow" effect, as if in a cave or well, or any empty area. It may be created electronically in an echo chamber or in a room by using several microphones.

echo chamber A separate studio, room, or tunnel that has a great deal of reverberation to create an echo effect. This method is outdated. In modern electronic applications variable delay lines are used.

ECO Ektachrome.

ecofilm Colloquial term for films and/or television film programs dealing with the environment.

ECS European Communications Satellite.

ECU Extreme close-up, also known as *XCU*.

ED Episodic dramatic program.

edge fogging   Light penetration along the edge of the film causing uneven exposure or "fog". It may be the result of loosely wound film or an insecurely closed magazine cover or film can. See also *fog*.

edge numbers   Also called **footage numbers** or **key numbers**. Latent numbers placed on the edge of the camera film stock at one foot intervals during the manufacturing process that will print in processing. They are visible on the positive film to facilitate matching of the scenes.

edge stripe   See *magnetic stripe*.

Edinburgh International Film Festival   A non-competitive film festival for independent films of general subjects held yearly in Edinburgh, Scotland (GB).

edit decision list (EDL)   Selected videotape shots logged in order for editing to be assembled in required continuity or kept for off-line editing at a later date.

editing   (1) Edit; videotape editing: elimination or replacement of individual shots to construct and build a program electronically. See *electronic editing*. (2) Live television: the director in the control room chooses the picture to be broadcast from among several camera monitors. (3) The assembly of film material, the cutting and elimination of unwanted portions of footage in the order of the script, and/or desired continuity. Film editing involves the cutting of the picture (rough cut, fine cut) and matching of the sound track. Film editing was introduced by the American director Edwin S. Porter ("The Great Train Robbery," 1903). See also *rough cut; fine cut;* and *sound track*.

editing bin   See *trim bin*.

editing in the camera   Filming sequences in direct continuity in order to minimize time-consuming and costly editing in post-production.

editing machine   An apparatus equipped with viewer, splicer, rewinds, and accessories, on which film can be reviewed running forward and backward, with short stops for cutting. Editing machines are usually combined with sound readers for synchronous process.

editing room   A light-controlled designated area where the videotape editing and/or film cutting takes place. The cutting of negative film is done in a separate, dust proof room. See also *negative assembly*.

editor   See (1) *Copy editor*. (2) *Assignment editor*. (3) *Videotape editing*. (4) *Film editor*.

editorial sync   Synchronous visual and sound markings on the picture and sound track to indicate the common starting point for editing.

edit pulse   A magnetic pulse marking the limit line between consecutive pictures, recorded on the control track of a transverse scan videotape recording.

edit start control (ESC)   A technique used to minimize picture distortion between segments where the tape is rolled back one second when the pause button is pressed.

EDL   Edit decision list.

EDTV   Extended definition television. See also *HDTV*.

educational broadcasting/film   Non-commercial broadcasting funded by individual or corporate donations, often by local, state, and federal governments, transmitting and/or producing classroom, direct teaching, adult, and general educational programs or films to schools, homes, and/or designated areas. See also *school broadcast*.

Educational Film Library Association (EFLA)   Former sponsor of the American Film Festival Blue Ribbon Awards and Emily Award. Now defunct (USA). See *National Educational Film & Video Festival*.

Eesti Radio   Radio Estonia.

Eesti TV   Tallin Eesti Televisioon (Estonia).

EFA   European Film Academy.

EFE   Agencia EFE (Spain).

effective isotropic radiated power (EIRP)   A measure indicating a satellite's transmitted signal strength.

effect lighting   Specially arranged lights, filters, patterns, and shades to achieve specific effects.

effect track   Sound effects recorded separately or after dialogue, music, or commentary, e.g. street noise, footsteps, etc.

effects (FX)   See *optical effects, sound effects, special effects*.

effects filter   Specially designed filter placed on the camera lens to create a specific visual effect like fog, a bright star, or diffusion.

effects library   A collection of catalogued and indexed sound effects, usually on records, tape, or film, for use by a broadcasting station or film studio.

EFLA   Educational Film Library Association.

EFS   Electronic filming system.

EFP   Electronic field production.

EG   Electronic graphics.

EGY   ITU country code for Egypt.

EHF   Extremely high frequency.

EI   Exposure index.

EIA   Electronic Industries Association (USA).

EIAJ   Electronic Industries Association of Japan.

Eidophor   Originally a large-screen monochrome television image projector. Also used in color by employing three monochrome projectors with additive color filters. Modern-day Eidophor systems employ light valve technology and no longer rely on CRT devices. See also *cathode ray tube projector* and *Triniscope*.

819-line   Television picture standard used in France and Monaco mandated by Charles DeGaulle. Now obsolete.

8mm   (1) Video 8; videotape format in 8mm. See also *hi 8*. (2) Solid state video camera designation. (3) Film format; amateur-type film with an 8mm gauge both in B&W and color. Now obsolete. See *Super 8*.

8XK   Call letters of the amateur radio telephone station operated by Dr. Frank Conrad in 1920 prior to the opening of KDKA, in Pittsburgh, Pennsylvania. See also *KDKA*.

80A filter   See *blue/2*.

85 and 85B filter   See *orange filter*.

EIRP   Effective isotropic radiated power.

EIRT   Ethikon Idhryma Radiophonias Tileorasseos (Greece).

E.I.S.   Europe Information Services (Belgium).

E = IT   See *reciprocity*.

E.J.   Electronic journalism.

Ektachrome (ECO)   American-made integral tri-pack 16mm reversal color film process and photographic film.

electrical power   (1) AC current supplied by regular wall outlets: 120V, 60Hz in the United States and Canada; 220V, 50Hz in most of the world, except Japan. (2) DC current supplied by batteries. Called *juice* in professional jargon. See also *battery*.

electromagnetic spectrum   See *spectrum*.

electromagnetic tape   See *magnetic tape*.

electromagnetic waves   Electromagnetic energy in the form of wave motion at the speed of light (299.792 km/s, or 186,282 mi/s).

electron beam   Electron gun; an electrode together with heater, cathode, and control grid that produces the electron beam in a television pick-up tube or CRT. Obsolete.

electron-beam recorder (EBR)   Videotape-to-film transfer effected by video signals while the film passes through a vacuum chamber. An improved method over the earlier kinescope recording. (Not successful.)

electronic animation   Animation carried out electronically with the use of computers vs. the conventional method on the animation stand. See also *computer generated images*-CGI.

electronic church (EC)   Religious broadcast programs and/or series. Commonly used term for televised church programs.

electronic editing, edit   Also called **videotape editing.** Computer-assisted editing of videotape programs by electronically erasing/inserting precisely timed program portions without physically cutting the tape.

electronic field production (EFP)   Remote video production, away from the studio, using portable equipment, cameras, recorders and accessory equipment.

electronic film transfer   See *kinescope recording.*

electronic graphics (EG)   Graphics, titling, and lettering done by an electronic (computerized) device to write, place, and arrange graphics and to adjust them directly on the screen.

electronic journalism (EJ)   Radio and television journalism by means of electronic equipment, such as tape and video recorders and cameras, to be broadcast over the air. See also *broadcast journalism.*

electronic news gathering (ENG)   News, news stories, interviews and news-related events recorded/photographed (gathered) by light-weight electronic equipment and sent back to the studio by either line-of-site microwave relays or via satellite. See also *satellite news gathering.*

electronic sports gathering (ESG)   Games, matches, tournaments and sporting events recorded and/or videotaped for either immediate broadcast via microwave relay or satellite (similar to electronic news gathering), or held to be broadcast at a later date.

electronic still store (ESS)   Generic term for an apparatus that stores a vast amount of color video images as single frames on computer disc files, which can then be easily accessed.

electronic transcription (ET)   Commercial recorded on tape or disc produced for radio and television.

electronic video recording (EVR)   Trade name for a video presentation using film and an electron-beam recorder for pick-up and a mechanism with flying spot scanner for production.

elephant ears   Colloquial term for a *flag*.

elevation   (1) Platform. (2) Floor plan design with indication of vertical planes. (3) The vertical angle of the satellite dish aimed toward a satellite.

ellipsoidal spotlight (ellipsoidal spot)   Also called **liko.** Studio light with spherical optics producing a "hard" beam.

ELS   Extreme long shot.

ELTA   Lithuanian News Agency.

EM   Exposure magnification.

EMBRATEL   Empresa Brasileira de Telecomunicações (Brazil).

emcee (MC)   Master of ceremony.

Emergency Broadcast System (EBS)   A group of announcements and procedures that may be activated in case of a national or regional emergency.

Eminent Artist and Merited Artist (Kiváló Művész and Érdemes Művész)   Recognition, award and title for outstanding contribution in film, television, music, and the arts. Given in either category by the government (Hungary).

Emmy Awards   Annual awards by the National Academy of Television Arts and Sciences, presented locally and nationally, in recognition of achievements in television (USA). National Emmy Awards are given in daytime programming, news and documentaries, sports, engineering, public service announcements, and community service. The Emmy statuette was designed by Louis McManus and the first award presentation was made on January 25, 1949 at the Hollywood Athletic Club.

The name Emmy is derived from Immy—the popular nickname of the Image Orthicon tube—and was suggested by Harry Lubcke, pioneering television engineer and former NATAS President (1949–50). See *National Academy of Television Arts and Sciences.* See also *Oscar.*

emulsion The light-sensitive gelatin and silver salt layer on the transparent cellulose base making up the film. The photographic images are recorded on this thin emulsion coating.

emulsion side The side of the film on which the light-sensitive emulsion is applied, as opposed to the *base*.

emulsion speed Also called **exposure index.** The rate of light sensitivity of the film (emulsion) expressed in numbers of ISO and/or DIN.

ENA Ethiopian News Agency.

encoder See *Colorplexer.*

encryption Signal scrambling technique designed to prevent program piracy. Descrambling requires a *decoder.*

end credits See *credits.*

endoscopic lens A specially designed thin, long lens for video cameras in the medical field, inserted into the body for crucial high-resolution pictures. Used especially with new HDTV applications. (From endoscope—the instrument for viewing internal parts of the human body.)

end slate Tail slate; slating shot taken (photographed) at the end of a scene instead at the beginning, with the slate board turned in an upside down position. See also *head slate.*

end test Laboratory test of a piece (end piece) of the exposed film for evaluation prior to the development of the entire reel.

ENG Electronic news gathering.

Engineering Achievement Award See *National Association of Broadcasters.*

engineering department Broadcast station or studio department with engineers in charge of all technical aspects. Headed by the chief engineer, who is directly responsible to the station manager (general manager).

enlargement See *optical enlargement.*

ENR Entreprise Nationale de Radio (Algeria).

ENTEL Empresa Nacional de Telecomunicaciones (Chile, Peru).

entertainment picture Theatrical film of comedy/drama produced for general audiences. See also *documentary, educational film, experimental film, industrial film.*

ENTV Entreprise National de Television (Algeria).

environmental sound   See *room tone.*

EOT   End of tape. See *automatic shut-off/*1.

epilogue   Concluding portion of a broadcast or film program.

episode   Part of a broadcast or filmed program serial presented at one performance.

EQA   ITU country code for Ecuador.

equalization   A technique used to bring the sound frequencies to recognized standards in quality recording and reproduction.

equalizer   Audio device inserted into channels of recording and reproduction to compensate for acoustical defects or to change frequency response. Used also for video signals sent over long distances by cables, which can result in loss of picture quality.

equal opportunity   Requirement by the Federal Communications Act to permit the use of broadcast time to legally qualified political candidates (USA).

equal time   Incorrect name for **equal opportunity.**

erase head   Also called **flying erase head** or **rotary erase head.** The magnetic assembly on a tape recorder that removes the previously recorded signals from the magnetic tape.

eraser   An electromagnetic device fitted with a special erase head to eliminate recorded sound and/or images from the sound and/or videotape. See also *bulk eraser* and *degausseur.*

ERP   Effective radiated power.

ERS (Emergency Radio Service)   Channel 9.

ERTT   Etablissement de la Radiodiffusion Télévision Tunisienne (Tunisia).

ES   (1) Emulsion speed. (2) Episodic sitcom.

ESC   Edit start control.

ESG   Electronic sports gathering.

ESPN   Entertainment and Sports Program Network (USA).

ESPRIT   European Strategic Programme for Research into Information Technology.

ESS   Electronic still store.

essential area   Also called **safe area** or **TV mask.** The area smaller than the film camera viewer and the film frame to allow for edges lost during television transmission as viewed on the home receiver. See also *full aperture.*

EST (1) Eastern Standard Time (USA). (2) ITU country code for Estonia.

establishing shot Also called **cover shot, opening shot,** or **orientation shot.** A photographic orientation shot that establishes the scene and brings in characteristic elements for the story; it is usually a long or wide-angle shot.

ET (1) Electronic transcription. (Obsolete for records and lacquer disks.) (2) Elliniki Tileorassi (Greece).

ETA Estonian Telegraphic Agency.

ETH ITU country code for Ethiopia.

ETS European Television Service.

ETV Educational television.

European Broadcasting Union (EBU) The association of European broadcasting services utilizing Euroradio and Eurovision (Geneva, Switzerland). See also *Euroradio/Eurovision.*

European Documentary Film Festival (Vue sur les Docs) International competition for feature, documentary and short films held in Marseilles (France).

European Film Academy (EFA) Founded after a meeting of filmmakers at the European Film Awards presentation in November 1988, EFA became part of the media program of the European Community (EC). Membership may be held as active, honorary, or associate. Among the Academy's aims are the exchange of ideas, the encouragement of solidarity and communication between European filmmakers, and the creation of public awareness of the modern European cinema.

EFA sponsors the Master Schools of Film, the European Summer Academy of Film and Media, various film symposia, and the annual Felix Award presentations; and publishes *Felix,* the Academy's biannual film magazine. See also *Felix.*

Europe Information Services (EIS) *European Report/Europolitique, La Lettre Européenne, La Lettre Sociale Européenne, Europe Environment, Europe Energie, Tech-Europe, Agromonde Service, Multinational Service, La Rapport Mensuel sur l'Europe* (Belgium).

Euroradio/Eurovision Network for the international exchange of national programs, operated by the EBU/UER—European Broadcasting Union/Union Européenne de Radiodiffusion (Geneva, Switzerland).

Eutelsat   European Telecommunications Satellite Organization (Paris, France).

event synchronizer   See *camera and event synchronizer.*

EVR   Electronic video recording.

exciter filter   See *ultraviolet photography.*

exciter lamp   A small lamp used in optical sound film projection to convert the modulation of light into audible sound.

exhibitor   The group or person in charge or owner of a cinema (movie) theater showing motion pictures.

existing light   The available light source (sunlight or house lights) without additional professional (studio) lamps. See also *artificial light.*

expansion   The widening of the volume range of a signal in an audio amplifier.

experimental film   A short film, very different from documentaries or theatrical pictures, seeking new elements in content, story, and/or production techniques.

experimental radio   (1) In 1920 the Physics Department of the University of Wisconsin operated an experimental radio station to broadcast weather and market reports. It was later called **WHA.** (2) In August 1920, Edward Wyllis Scripps (1854–1926) of the *Detroit News* started his experimental radio station, now called **WWJ**. See also *8XK* and *KDKA.*

exposed film   Film subjected to light, photographed onto through the camera lens, but not yet developed in the processing laboratory.

exposure   The process of exposing the motion picture film to light in the camera or printer in order to produce a latent image on the emulsion coating. The exposure is determined by the time and the degree of illumination.

exposure control   Control exercised both in the filming and printing (laboratory) processes to maintain correct exposure and to ensure uniformity.

exposure index   See *emulsion speed.*

exposure latitude   The range of exposure between an underexposed and an overexposed film that produces an acceptable image.

exposure meter   Small light-sensitive device that determines the light flux indicating the required f-stop and exposure time. See also *incident light meter* and *reflected light meter.*

exposure setting   The selection of the lens opening (f-number) and shutter speed to expose the motion picture film.

exposure time   The actual time, a fraction of a second, that each motion picture film frame is exposed to light. 1/50 of a second at 24 frames per second; 25 fps in the European-Continental television system.

EXT   Exterior.

Extended Definition Television (EDTV)   A transition between the customary existing television set and high definition television (HDTV).

exterior (EXT)   Scene or part of a television show or film taking place and/or photographed outdoors. See also *interior.*

exterior lot   See *lot.*

external broadcasting   Broadcasting services run by governments with programming directed toward foreign countries.

extra   Also called **walk-on.** A person engaged by the studio or production company to perform minor parts or to be part of a crowd in a film or show.

extreme close-up (ECU)   Tight photographic shot taken at close range or by a narrow-angle telephoto lens, showing a very close picture or detail of the subject. See also *close-up* and *macro cinematography.*

extreme long shot (ELS)   An extremely wide angle shot taken at long range and/or by a short focal-length lens showing broad, vast distances. See also *fisheye lens* and *wide-angle shot.*

extremely high frequency (EHF)   30,000-300,000 MHz; with a wavelength of 1cm.

eyebrow   A small metal flag attached to the matte box on the film camera to shade the lens. See also *sunshade.*

eye cup   Soft rubber cup attached to the viewfinder to shield the eye from side light.

eye level   Also called **eye line.** A shot taken by the camera raised or lowered to the same horizontal plane as the eye.

eye light   See *camera light.*

eye piece   Small lens or lenses at the eye-end of a camera's viewfinder and adjustable to the camera operator's eye. See also *viewfinder.*

f   See *f-stop.*

F   ITU country code for France.

facilities man/woman   Floor man/woman, grip, or stage hand.

facsimile (FAX)   The transmission of printed material or still pictures over the airwaves.

FACT   Federation Against Copyright Theft (GB).

fact sheet   Also called **run-down sheet.** The list of particular items that must be included in a broadcast program. Often used for non-scripted (ad-lib) advertisements.

FAD   First assistant director.

fade   (1) An increase or decrease in sound volume. (2) The gradual appearance or disappearance of a picture from black or to black— fade in, fade out. See *dissolve.* See also *cross fade.* (3) **"Fade sound and picture"**—command given by the television director in the control room to end the program or show.

fader   The control (knob) for sound or vision on the mixing panel (console). Often a **potentiometer** (abbreviated as **pot**).

fading   Intensity variations of a radio signal resulting from changes in the transmission.

FAM   The Family Channel cable television network (USA).

FANTASPORTO   Oporto International Film Festival; annual film festival with emphasis on new directors, held in Oporto, Portugal.

fast film   Motion picture film with high ISO rating and high sensitivity to light thus requiring less illumination. See also *slow speed film.*

fast forward   Advancing the tape forward rapidly to reach a desired selection. The opposite of *fast rewind.*

fast lens   A camera lens with a large aperture and small f-stop with excellent light collecting properties. See also *slow lens.*

fast-mo   Colloquial for *fast motion*. See also *slo-mo*.

fast motion   Running the film in the camera slower than normal speed while photographing a scene. When the film is projected at a standard rate, the motion appears faster, accelerated. The opposite of *slow motion*.

fast rewind   See *rewind*.

FAX   See *facsimile*.

FCC   Federal Communications Commission (USA).

feature film   A theatrical motion picture produced for entertainment and having a fictional story line. The average running time is 3,000 feet (900 meters) in 16mm format or about 8,000 feet (2,400 meters) in 35mm format. Feature films are usually shown in cinema (movie) houses.

FEBA   Far East (Missionary) Broadcasting Association (Seychelles).

Federal Communications Commission (FCC)   Independent agency of the U.S. government, established in 1934 as a successor to the Federal Radio Commission of 1927. The FCC regulates radio, television, wire and cable communications in the United States and to and from foreign countries. The seven FCC Commissioners are appointed by the President with the approval of the Senate. Each appointment is for a seven-year term. The commissioners are responsible to the Congress.

   The Commission deals with rate increases for telephone and telegraph services; allocates broadcast bands and frequencies; assigns call letters, power, and specific frequencies; issues station and operator licenses; monitors broadcasts for possible violations; and provides assistance in distress for air and sea traffic.

Federal Radio Commission (FRC)   Established by the U.S. Congress in 1927 for the regulation of radio communication to the public. In 1934, its successor, the Federal Communications Commission, was established. See *Federal Communications Commission*.

Federation Internationale des Archives du Film (FIAF)   International Federation of Film Archives; international organization that maintains data on film archives throughout the world and organizes meetings for archivists (Paris, France).

feed   Feeding; transmission of radio or television signals from one program source to another, from a network to individual stations, or from a remote (OB) broadcast source to the studio.

feedback   (1) Disturbing, often piercing howl or squeal, over amplified sound returned from a loudspeaker back to its source (microphone). See also *Larsen effect.* (2) Audience reaction at a live performance as sensed by the performer.

feed reel   See *supply reel.*

Felix   The European film award, presented annually by the European Film Academy and sponsored/financed by national public funds. The 57cm (appr. 22.5 in.) Felix award statue was designed by Prof. Markus Lupertz and is promoted as the European equivalent of the American "Oscar." See *European Film Academy.*

FENA   Far Eastern News Agency (Republic of China).

Ferraniacolor   Italian-made negative/positive integral tri-pack color film process.

ferrous oxide   See *oxide.*

FESPACO   Festival Pan Africaine de Cinema de Ouagadougou; yearly Pan African film and television festival for Africans and filmmakers of African origin, held in Ouagadougou (Burkina Faso).

Festival des Films du Monde—Montréal   International festival emphasizing film premiers, held annually in Montreal (Canada).

Festival International du Film—Cannes   Highly recognized International Film Festival held in Cannes in the French Riviera since 1946. The annual event takes place in the spring and its principal award, the Palme d'Or (The Golden Palm), is given in both the feature and short film categories. Other awards are the Grand Award of the Jury, awards for the Best Male acting role, best Female acting role, best Production design, the Special Award of the Jury, and Camera d'Or (The Golden Camera) for cinematography. The Festival also serves as an international meeting place for the promotion and distribution of newly released films.

Festival of Festivals   See *Toronto International Film Festival.*

FF   Film festival.

FIAF   Federation Internationale des Archives du Film (France).

fiber-optic   A hair-thin glass fiber with very large capacity and practically no interference, used increasingly in telecommunication of audio, video, and data signals that have been converted to light impulses.

fiction   Literary work (book, story, or script) as a result of imagination rather than facts or actual events; i.e. fictional film.

fidelity   The degree of accuracy in sound reproduction.

field   (1) One-half of the complete television scanning cycle. Two interlaced fields are necessary for a picture—60-field/sec. in the United States and 50-field/sec. in Europe and most of the world. (2) See *depth of field.*

field camera   Portable camera.

field lens   See *telephoto lens.*

field of coverage   Area covered by the camera (lens).

field of view   Also called **angle of view.** The area seen by a respective lens. See also *angle.*

field of vision   Field of coverage.

field pick-up   Outside broadcast (OB) or remote (location) broadcast.

field rate converter   Converter designed to change the television field rate e.g. 50Hz to 60Hz and vice versa, by using a helical scan recorder for storing information.

FIFARC   Festival International du Film d'Architecture, d'Urbanisme et d'Environnement; film festival for architectural, urbanism and environmental subjects held biannually in Bordeaux, France.

FIFART   Festival International du Film d'Art; international festival for art, crafts, animation, and documentary films held in Lausanne, Switzerland.

15 ips   See *tape speed.*

15/16 ips   Magnetic tape speed for "Books for the Blind."

50-field   See *625-line/50-field.*

50 ft   Film length standard—15 meters.

50.8mm   Magnetic tape width standard—2 inches.

FII   Film Institute of India.

fill   Broadcast program material prepared in advance for filling time in case the show runs short (usually after special events and sportscasts).

fill light   Also called **filler light.** Fill-in light; soft studio light to reduce shadow effects and high contrast areas. See also *booster/2.*

film   See *motion picture film.*

film archive   An organized repository for films and film programs. See also *Federation Internationale des Archives du Film.*

film base   See *base/2.*

Film Board of Canada   See *National Film Board of Canada.*

film camera   See *camera/2*.

film can   Metal or plastic container for storing and shipping film material.

film car   Cinema van.

film cement   See *cement*.

film chain   Television film and slide projection chain and the room where the equipment is housed. See also *film island* and *telecine*.

film cleaning   See *cleaning*.

film clip   See *clip/2*.

film development   See *processing*.

film editor   Member of the production team who assembles, selects and composes film sequences in stages, then cuts and matches the accompanying soundtrack into a complete program.

film festival   National and/or international competitive, non-competitive, or invitational film and/or video presentation of recent productions of various length and subjects, held usually at annual or bi-annual intervals. See entries of individual and specific film, television, or video festivals.

film frame rate   See *25 fps* and *24 fps*.

film gate   See *gate/1*.

film island   Film chain or telecine; consist of one or two film projectors, slide projectors, a multiplexer, and a television video camera or scanner device.

film length   (1) Running time of a filmed program. (2) The physical length of the film on reel or spool, expressed in feet and/or meters.

"Filmless" camera   Colloquial for digital photography device (camera) now under development at Sony.

film librarian   Person in charge of receiving, cataloging, selecting, and handling library stock footage, film, or video collections.

film library   The room or area where library stock footage, film, and/or videotape collections are kept, catalogued, stored and circulated.

film loop   (1) Continuous band or small reel of tape or film spliced together head to tail and used for uninterrupted projection and -film screening. (2) Loose part of the film wound through sprocket-equipped guides to give it substantial play (in order to prevent breakage) as the film pases through the gate of the camera or projector.

film magazine   See *magazine*.

film number   Number(s) identifying the various types of films.

film perforation   See *perforation*.

film processing   See *processing*.

film rack   See *rack*.

film rate   See *25 fps* and *24 fps*.

film rejects   See *out takes*.

film release   See *general release* and *release/4*.

film rupture   The breakage of the film in the camera, projector, or processing machines during transport.

film slitter   Machine designed to cut film down the center, i.e. 16mm to 8mm, or 35mm to 17.5mm.

film speed   See *emulsion spèed* and *speed/3*.

film speed indicator   See *footage counter*.

film splicer   See *splicer*.

film standard   Accepted measures of film width, types, lengths, ratios, and dimensions commonly used in professional practice. See *Super 8, 16mm film, Super 16, 35mm, 65mm, 70mm film; 100', 400'*, etc.

film star   See *star*.

film storage   See *vault*.

film strip   A strip of 35mm film with single pictures shown in a special (film strip) projector. Film strips can be silent with subtitles or can be accompanied with corresponding sound on disc or tape. Used extensively in education, training, meetings, conventions, and marketing.

film studio   See *studio*.

film threading   See *threading*.

film track   Film guiding component in the camera, projector, or developing machines.

film transfer   See *kinescope recording*.

film unit   A film production team at a broadcast station, e.g. documentary film unit.

film viewer   See *viewer*.

film window   See *gate/*1.

filter   (1) An electronic device designed to eliminate or accept a certain range of frequencies, thus modifying the quality of the transmitted signal. (2) A toned, transparent (glass or gelatin) material placed in front of a camera lens to change the character of light and achieve specific photographic values. See *blue filter, contrast filter, correction filter, haze filter, ND, orange filter, Polarizing filter*. See also *daylight conversion/*2.

filter factor   A factor expressed in numbers indicating the required increase of exposure due to the loss of light absorbed by the filter.

final cut   The completed, edited workprint.

final print   Release print.

finder   See *viewfinder* or *director's finder*.

fine cut   The final stage of the workprint approaching completion when through a series of rough cuts all unnecessary elements (scenes, footage, frames) have been taken out. See also *rough cut*.

fine grain   A film emulsion made up of very small (fine) particles and used for black-and-white duplicate negative stock.

FINTEL   Fiji International Telecommunications, Ltd.

FIPA   Festival International des Programmes Audiovisuelle; international audiovisual festival for television, entertainment and documentary programs held annually in Paris (France).

first generation   The original film or tape (master). See also *master/*4.

fishbone pattern   A reception interference to television channels causing visual distortion and moving lines. Caused by close proximity to unwanted sources of radio frequency interference—RFI.

fisheye lens   An extreme wide-angle lens capable of covering a very large (far beyond the norm) area. It may render a highly distorted view in close situations (CU).

fishpole   A lightweight bamboo, aluminum, or fiberglass pole, similar to a fishing rod, on which a microphone is fastened and handheld over the scene or location, or in a studio with participating live audiences.

5-inch reel   Diameter for magnetic tape reel, equals 12.7cm. See also *10.5-inch reel* and *14-inch reel*.

525-line/60-field   North American television picture standard. See also *625-line/50-field*.

5400K   Average color temperature of daylight at noon on the Kelvin scale. 2000K at sunrise or sunset, and 5800K in midsummer. See also *3200K*.

fixed focal-length lens   A lens, usually wide-angle, with only an adjustable f-stop, while the distance setting, thus the focus, is fixed.

fixed station   Colloquial for fixed location base station for two-way radio systems.

fixing   Fixing bath; film laboratory process following development to stabilize the film.

FJI   ITU country code for Fiji.

flag   A small, black-surfaced and square-shaped sheet or card, usually mounted on a stand or on the matte box to shade camera lenses. See also *french flag*.

Flaherty Documentary Award   A Lloyds Bank BAFTA Production Awards presentation for the best documentary film on television (GB). See also *British Academy Awards*.

flange   A plastic or metal disc against which the film is wound or rewound on a core.

flaps   See *barn doors*.

flare   (1) Light reflection caused by shiny surfaces. (2) Unintentional exposure of the film or portion thereof caused by light leaks, faulty lens element, etc.

flare spot   See *hot spot*.

flash   (1) A very short sequence or shot. (2) Laboratory technique to lighten film sequences. (3) See *news flash*.

flashback   Return of the action to a previously shown scene or earlier time period.

flashing   See *pre-flashing*.

flat   (1) A piece of standing scenery. (2) Uninteresting scene or action. (3) Insufficient contrast in picture reproduction. (4) Flat rate, flat fee.

flat light   Even, overall soft lighting with no modeling or highlights.

flat plate antenna   A specially constructed antenna utilizing a great number of reflective elements as opposed to a dish reflector.

flat rate   Also called **flat fee.** Fixed payment to actors and production personnel.

flat response   A desired frequency response in audio work rated plus or minus 3 dB from 50 to 15,000Hz.

flicker   (1) Light intensity changes in projection caused either by the shutter interference or a slower pace of projection disrupting the illusion of continuous movement (persistence of vision). (2) Slang for an entertainment type film.

flies   Center-weighted pulleys to hoist scenery.

flip card   Caption or title cards of identical sizes that are changed (flipped) off at a pace set by the director.

flip-over   Also called **flip over wipe.** Optical effect rotating the image on the screen horizontally or vertically to its reverse side.

FLK   ITU country code for the Falkland Islands.

float   Irregular movement of the film, unsteady image in the film camera or projector.

flood   Floodlight; non-directional, diffused light used for general illumination and for fill light.

floor   The stage; the television or film studio, where the production takes place.

floor director   Floor manager.

floor manager   The stage manager of the television studio; the essential link between the control room and the studio; the director's right-hand man/woman on the floor. He/she is in charge of the studio and all activities during the production.

floor men/women   A team of "hands" responsible for erecting, changing or striking sets, set dressing, and other similar jobs in the studio. See also *grip* or *stage hands.*

floor mixer   See *recordist* (British term).

floor plan   A scaled studio "map", the stage plan, proportionately indicating the sets and properties in relation to the studio layout.

floor stand   Microphone support with adjustable height used primarily in variety and musical shows. See also *boom* or *table stand.*

fluff   Also called **beard.** A mistake, an error made by a talent (actor) during performance.

fluid head   Also called **liquid head.** A relatively lightweight tripod head without gears, employing a thin film or liquid to ensure smooth movement. See also *friction head* and *geared head.*

fluorescent light See *cold light.*

fluting Swelling of the outside edge of the film on a roll caused by excessive humidity. See also *buckling.*

flutter A fault, a cyclical frequency deviation in sound recording or reproduction. See also *wow.*

flux See *luminous flux.*

fly Scenery hanging above the set.

flyaway Movable satellite uplink unit.

flying erase head See *erase head.*

FM Frequency modulation.

FM stereo See *multiplexing.*

FNL ITU country code for Finland.

FNN Financial News Network (previous name for CNBC) (USA).

f-number Also called **f-stop.** Calibration expressing the relative opening of the lens; the ratio between the diameter of the aperture and the focal length of the lens. Smaller f-number means larger opening. See also *t-number, t-stop.*

focal distance Focal length.

focal length (focal distance) (1) Lens measurement in millimeters or inches indicating the distance from the optical center of the lens, focused at infinity, to the surface of the camera tube in television. (2) From the optical center of the lens, at infinity setting, to the film emulsion surface in motion pictures. See also *long focal-length lens* and *short focal-length lens.*

focal plane The area in space on which an image passing through the lens, perpendicular to its optical axis, is formed.

focal point The focal point of rays in general; a point of the lens where parallel rays converge (real focus), or from which the passing rays diverge (virtual focus).

focus (1) A principal point where sound waves converge. (2) The point at which the subject must be positioned so that the image in the lens will be sharp and well defined.

focusing Moving the camera lens to or away from the camera image sensor (television) or emulsion plane (film) by rotating the focusing ring until the image comes into sharp focus ("bring to focus").

focusing light Studio and stage light with an adjustable bulb that can be moved back and forth to facilitate focusing.

focusing ring Also called **sleeve.** A grooved ring on the lens barrel which is periodically adjusted for focusing or de-focusing.

focus puller Operator, member of the film camera crew, who adjusts the lens for focusing. (British term). In the United States the job is usually done by the first assistant cameraman/woman.

fog/fogging (1) Accidental exposure of film caused by light streak. (2) A result of overdevelopment. (3) Old, outdated film. (4) Insufficient storage of the film in a hot, humid area. See also *edge fogging.*

fog filter A diffusing lens filter that gives the effect of static fog.

fog gun A small, portable fog machine operated with a heating unit.

fog machine Device of various types and sizes for creating fog on the set. Fog is created most frequently by a combination of hot water, dry ice, and a fan (caution must be exercised when using a fog machine, for it may cause audio problems).

fold back Fold back system; the technique of using a special portion of an audio mixing console to send a special band or mix of audio back to the studio for the performers to hear for cuing or lip sync purposes. The special mix is "folded back" to the cast in the studio.

follow focus A technique to adjust the focusing sleeve on the lens so that a moving object is kept in constant clear focus. See also *stop pull.*

follow shot A shot taken with a moving camera, moving with or along the subject.

follow spot A powerful spotlight that can be moved to keep a performer in the same spot circle.

foot British/American linear measure, 12 inches. Equals .304 meters, or 30.48 centimeters.

footage (1) Length of film measured in feet rather than meters. The term is also used for noting "good footage" or "bad footage." (2) Colloquial for the film itself.

footage counter Sprocketed metering device showing the film footage length run through it. Footage counters are found in cameras and editing and processing equipment. Editors use a single unit counter.

footage number See *edge numbers.*

foot candle Unit of light intensity; the amount of light produced on a sphere by a candle one foot away. Now obsolete. See *candela.*

foot lambert   Unit of luminance; equals one candela per square foot of an emitting source or radiating surface.

footlights   A row of lights along the front of a stage or studio set.

footprint   The geographical coverage area of a satellite; the area of the earth within which a satellite signal can be best received.

foot switch   A mechanically or electrically operated foot pedal that starts and/or stops recording/playback equipment.

forced development   Also called **push processing.** Forced processing; laboratory process to increase the contrast in underexposed camera film, achieved by the increase of temperature and time in development.

foreground   The part of the studio stage or set close to the camera(s).

format   (1) Scripted elements of program concept, an established pattern of presentation. (2) Size or aspect ratio of a standard television picture and/or the motion picture frame.

45 rpm   See *recording speed.*

45%   Percentage of relative humidity at 72° F (22.2° C); the ideal condition for film or tape storage. See also *vault.*

4As   AAAA; American Association of Advertising Agencies.

four channel   Also called **quadraphonic.** Stereophonic recording and reproduction system using four channels.

400 ft   Standard film length, equals 122 meters.

405-line/50-field   UHF television picture standard formerly used in Great Britain, now obsolete.

4.76 cm/s   Magnetic tape speed, equals 1⅞ ips.

14-inch reel   Magnetic tape reel diameter, equals 35.4 cm. See also *5-inch reel* and *10.5-inch* reel.

FOX   Fox Television cable network (USA).

fpm   Feet per minute.

fps   Frames per second.

frame   (1) Frame frequency; a single complete television picture (two fields); 30 fps—USA; 25 fps—Europe and most of the world. See also *field.* (2) A single picture on the motion picture film. (3) To compose a picture, "to frame."

frame counter   A dial along the footage counter indicating (counting) the frames of a respective film format being rolled through.

framer A mechanical device in a camera viewer or projector, used to adjust frame lines. See *framing knob/lever*.

frame store See *electronic still store*.

"frame up" Call to the camera operator to properly center the subject, scene, or action.

framing knob/lever Component part of an image framer in a viewer, editing machine, or projector.

FRANCE TELECOM The French national satellite network in operation since 1984 (Paris, France).

FRC The former Federal Radio Commission, now FCC (USA). See *Federal Communications Commission*.

FRCN Federal Radio Corporation of Nigeria.

Freedom Foundation Award Annual award given to radio and television stations featuring creative work in advocacy of better understanding of America.

free-lance Independent, non-staff professionals doing work assignments for a studio or production house.

freeze (1) Production arrangements approved. (2) Actor or talent who went blank and forgot the lines.

freeze frame Also called **hold frame** or **still frame.** Freezing; the holding or repeat printing of a single frame of action for added effect or attention. In television it is achieved electronically, in film it is done by laboratory technique.

french flag A small metal plate fixed to an adjustable arm and placed in front of a light to shade certain areas. See also *flag* and *gobo*.

frequency The number of complete oscillations per second of an electromagnetic wave. The unit of frequency is hertz (Hz).

frequency band See *band/1*.

frequency modulation (FM) The frequency variation of the carrier wave while its amplitude remains constant. It provides a method of signal transmission free from static interference. Invented in 1933 by Edwin Howard Armstrong (1890–1954), American electrical engineer. See also *amplitude modulation* and *modulation*.

frequency response The ability of the audio or video system to properly record or reproduce the entire original frequency range of signals without too severe loss or attenuation of the low- or high-end signals, which are the most unstable.

Fresnel lens   A heat-tolerant lens made lighter and thinner by a stepped surface (echelon), used in spotlights.

FRI   ITU country code for Faroe Islands.

friction head   A camera tripod head that ensures smooth movement with a sliding (friction) surface. See also *fluid head* and *gyro head.*

fringing   Imperfect registration overlapping two or more images on film; especially noticeable in color film.

"from the top"   Start from the beginning; indication voiced at rehearsals.

front end   See *down converter.*

front projection   Most common form of film projection whereby the image is thrown directly onto a reflecting (non-translucent) screen in front of (opposite) the projector. See also *back projection.*

fry   See *scratch.*

FS   Full shot.

f-stop   f-number.

ft   Foot measure, equals 30.48cm or .3048m.

FTN   Fuji Television Network (Japan).

Fujicolor   A negative/positive integral tri-pack color film process made in Japan.

full aperture   Also called **silent frame.** The full size of the aperture in the camera or projector mask.

full coat   Magnetic film, usually in 16mm or 35mm formats, fully coated with iron oxide, used for multiple track recording. See *magnetic film.*

full frame   Full aperture. See also *essential area.*

full shot (FS)   (1) Loosely used term indicating a shot that covers the entire subject area or scene. (2) A shot taken with the subject filling the frame.

fully scripted   A completed script giving all audio and video information. See *script* and *shooting script.*

funnel   See *snoot.*

FX   Abbreviation for **effects.** See also *SFX.*

G

G (1) ITU country code for the United Kingdom of Great Britain and Northern Ireland. (2) General audiences; film rating symbol.

GAB ITU country code for Gabon.

Gabriel Award Award presentation in eleven categories to radio and television stations for programming that promotes a value-centered vision of humanity. Sponsored by the National Catholic Association of Broadcasters and Communicators (USA). See also *NCABC* and *UNDA*.

Gaevert See *Agfacolor (Agfa-Gaevert)*.

gaffer The chief electrician in the film studio or on the set. He/she works with and is responsible to the director of photography.

gaffer grip Lightweight but heavy-duty spring-loaded aluminum grip for hanging lighting equipment in confined areas.

gaffer tape Also called **camera tape.** Two-inch (50mm) wide, aluminum colored, strong pressure-sensitive adhesive tape used in set dressing and rigging.

gain Rate of amplification of an electronic system, level of amplified sound, brightness of the picture. See also *riding gain*.

Galaxy Domestic satellite operated by the Hughes Corporation (USA).

gamma Degree of contrast of a photographic image. A term used in video to describe the center portion of a 10-step gray scale and the adjustments that can be made to alter the video contrast ratio.

gang synchronizer See *synchronizer/2*.

gantries See *catwalk*.

gap The tiny opening between the poles of tape heads, measured usually in microns.

gate (1) Camera and projector aperture with a hinged unit that holds the film in place while it is stopped briefly for light exposure. (2) A small gap between two pieces of film felt by a splicer as a result of non-precise alignment.

gator grip   A light socket-swivel type clamp for flat table edges or poles to support lightweight lamps.

gauge   The width of standard film or tape.

gauze   Thin, transparent fabric made of cotton, silk, or wire used as a light diffuser. See *scrim*.

Gavel Award   Annual award by the American Bar Association given for superior law-related work in the mass media.

GBC   (1) Gibraltar Broadcasting Corporation. (2) Ghana Broadcasting Corporation.

G.C.   Geographical coordinate.

GDL   ITU country code for Guadeloupe.

geared head   A crank handle-operated tripod head with a geared device to achieve smooth horizontal and vertical movements. See also *fluid head* and *friction head*.

Geijutsu Sakuhin Sho   Annual awards by Japan's Ministry of Education for best television drama, television documentary, radio drama, radio documentary, and record production. The prestigious awards presentation takes place in March. See also *Nippon Academy Awards*.

gel   See *gelatin filter*.

gelatin   (1) Purified substance of complex protein used as basic emulsion material in the manufacture of photographic film. (2) Filter used to change the color or quality of light.

gelatin filter   Also called **gel** and/or **jelly.** A translucent sheet, or a camera or light filter placed on a stand or in front of the camera or light to reduce the intensity of light, correct color, or alter (distort) light in order to achieve an artistic effect.

Gemini Awards   Yearly awards presentation by the Academy of Canadian Cinema and Television celebrating achievements in English-language Canadian television. The Gemini Award sculpture was created in 1986 by designer Scott Thornley. See *Academy of Canadian Cinema and Television*. See also *Prix Gemeaux* and *Genie Awards*.

general release   The promotion, distribution and exhibition of a film (usually a feature) in movie theaters.

generation   See *first generation*.

generation loss   Deterioration, loss of sound and/or picture quality due to duplication.

generator   Self-contained power supply trailer or truck generating electrical current. Used extensively on location.

Genie Awards   Annual awards presentation by the Academy of Canadian Cinema and Television honoring excellence in Canadian cinema. See *Academy of Canadian Cinema and Television*. See also *Gemini Awards* and *Prix Gemeaux*.

gen-lock   A feature of professional-grade sync pulse generators which locks to provide unison scan from all (various) sources and to prevent unwanted rolling of television pictures.

genre   French term indicating type of film and broadcast programs either by content or by their specific audience, like comedy/variety, soap opera, daytime serial, or children's program.

GEO   ITU country code for Georgia.

geostationary orbit/geosynchronous orbit   See *synchronous orbit*.

GHA   ITU country code for Ghana.

ghost   Double image on the television receiver caused by the reflection of signals off tall buildings, mountains, or approaching planes near the receiving antenna.

GHz   Gigahertz—1,000,000,000 Hz.

GIB   ITU country code for Gibraltar.

gimbal tripod   Camera mount with a free-swinging weighted pendulum.

gimmick   A new approach, a change, a trick in a presentation.

gizmo   Generic term used for devices or equipment parts in lieu of their proper term or when one is unable to remember the actual name.

Glasses cam   See *video sunglasses*.

glass shot   Also called **glasswork.** Trick photography in which a partially painted glass is photographed together with the main action. Note: It is not the same as *mirror shot*.

glitch   (1) Term generally used to denote a problem or unexpected difficulty. (2) A brief disturbance in a videotaped image often caused by a wrinkled or edge-damaged videotape or an objectionable number of physical drop-outs.

glob   Global beam; satellite footprint covering the earth's surface.

gm   Gram.

GMB   ITU country code for Gambia.

GMT Greenwich Mean Time. See also *UTC*.

GNA Ghana News Service.

GNB ITU country code for Guinea Bissau.

GNE ITU country code for Equatorial Guinea.

gobo (1) A light shield made of a black screen or wood, mounted on a stand close to a lamp. See also *french flag*. (2) A portable, off-camera, sound-absorbing screen or wall.

gofer (go-fer) Colloquial term for entry-level production assistant, "Go for . . . ".

Golden Eagle Awards Annual film and video competition award presentation in some 20 categories by CINE. See *Council on International Nontheatrical Events*.

Golden Gate Award Presentation to winners of the annual San Francisco International Film Festival (USA). See *San Francisco International Film Festival*.

Golden Globe Award by the Hollywood Foreign Press Association for the best judged film or television production in comedy and drama. The annual awards presentation is held in January (USA).

Golden Palm Palme d'Or; award of the renowned annual Cannes Film Festival held in France. See *Festival International du Film-Cannes*.

good footage Acceptable, good camera footage. See also *N.G.*

good take A satisfactory recording or film shot.

Gorizont (Horizon) Russian fixed satellite system.

GOST Film emulsion (speed) system used in the former USSR.

Göteborg Film Festival International festival for 16mm, 35mm and 70mm films previously not exhibited in Sweden. Held in Göteborg (Sweden).

governor motor Also called **constant speed motor.** A type of film camera motor used in double system (synchronous sound) filming that maintains an actual constant speed at all times.

Goya Name of the national film awards presented by the National Academy of Cinematographic Arts and Sciences of Spain.

grading See *timing/2*.

grain Particles, a combination of metallic silver and silver halide crystals, exposed and developed to form an image on the film.

graininess   Random pattern of grain particles (in clumps) that are noticeable to different degrees during film projection. Films with higher emulsion speeds have more graininess.

gram (gm)   Unit of mass, 1/1000th kilogram. Equals 0.0352 ounce.

graphics   Lettering, logos, sketches, design; artwork in general.

gray scale (gray card)   Standard B&W television test card with stepped tonal graduation from black to white for the control of picture tonal quality. Maximum step: 10; satisfactory gray scale: 7; poor: below 5. (Earlier some studios used a scale that was toned green, since the green reproduced a better gray on monochrome television.) Recent EDTV and HDTV systems provide for 15 or more gray-scale steps.

GRC   ITU country code for Greece.

GRD   ITU country code for Grenada.

grease pencil   See *marking pencil.*

green   Additive primary color.

green film print   Release print fresh out of the processing laboratory, with excessive moisture on it.

grid   Overhead parallel bar framework on which lights and scenery are hung.

grip   Floor man/woman or stagehand who works closely with the camera crew.

GRL   ITU country code for Greenland.

gros plan   French for *close-up (CU).*

ground   The part or point of an electrical system with zero voltage; the part connected to an earth ground.

ground film   A piece of opaque film used for image focusing in specific cameras, especially in boresight viewing.

ground glass   A piece of focusing glass with a finely ground surface built into the camera viewfinder to facilitate framing, composition, and focusing.

ground row   A strip of often natural material placed low to hide junction of scenery and footlights in order to achieve a realistic effect.

ground station   Television receive-only antenna (TRVO).

ground tripod   See *high hat.*

ground waves Electromagnetic waves produced by transmitting aerials that travel along the ground. Ground waves carry long and medium wavelength transmission and cover only short distances, AM, MF. See also *direct wave, sky wave,* and *microwaves.*

Grover Cobb Award See *National Association of Broadcasters.*

GSAP Gun sight aiming point camera (16mm).

GTM ITU country code for Guatemala.

GUATEL Empresa Guatemalteca de Telecomunicaciones. (Guatemala).

GUF ITU country code for French Guiana.

GUI ITU country code for Guinea.

guide track Sound track synchronized and recorded simultaneously with the picture track, used only as a guide and not for reproduction purposes.

Guild of Television Producers and Directors, The A former professional organization that later, with the British Film Academy, formed the Society of Film and Television Arts, then became The British Academy of Film and Television Arts (GB). See *The British Academy of Film and Television Arts.*

guillotine splicer Device used in film cutting or editing for faster operations. It uses non-perforated tape which it automatically cuts and perforates in one step. The cutters may be straight, diagonal, or both.

Guldbaggen Golden Beetle; annual Swedish film award.

GUM ITU country code for Guam.

gun mike See *shotgun microphone.*

gun sight aiming point camera (GSAP) A lightweight 16mm film camera used in skiing, sky diving, and auto racing for "point-of-view" effect.

GUY ITU country code for Guyana.

gyro head Gyroscopic head; a camera support tripod head incorporating a gear train and heavy flywheel to ensure smooth movement.

gyro zoom Trade name of an image stabilizer lens used in motor vehicles, helicopters, airplanes, boats, or hand/shoulder held camera shots to eliminate image vibration.

H

h   Hour.

H   (1) Horizontal (polarization). (2) Henry; the henry. (3) Horizontal line (in NTSC: 63.5 microseconds).

HA   Haber Ajansi (Turkey).

HAB   Hawaiian Association of Broadcasters.

halation   A halo-like ghost image appearing around a light subject on the film emulsion caused by light reflection of the film base. See *anti-halation backing*.

1/2-inch   Standard magnetic tape width standard, equals 12.7mm.

half loop/wrap   Magnetic tape wound halfway around the drum in 180°. See also *alpha loop* and *omega loop*.

half track   Also called **dual track.** A ¼-inch audio tape recording with two tracks, and with a wider guard track in between to minimize crosstalk.

halide   Silver halide; a light-sensitive chemical compound of a halogen with other elements that forms the emulsion part of film.

halitosis   Colloquial for white spot on the film, a result of dirt on the negative.

halo effect   Ghost image around a light subject on the television screen, similar to **halation.**

ham   (1) A licensed amateur shortwave radio operator. (2) A (self-centered) actor or performer who exaggerates and overacts.

hand crank   (1) A mechanism for winding the manual film camera (spring) motor. (2) A small mechanism for rewinding film in a camera for lap dissolve or double exposure. (3) A geared crank handle on zoom lenses that allows for smooth movement.

hand-held   A scene or shot photographed without a tripod or fixed mount by holding the camera and/or mike in hand or on a shoulder pod. See also *Steadicam.*™

handle (1) Overlap video and audio material on either side of a video-tape edit used for optical effects, like a dissolve or swish. (2) CB slang for personal ID.

hand props Small, personal property handled by the actors and performers.

hand signals Visual cues and signals given by the director, floor manager, or other designated production personnel in the studio or on the set where voiced commands would be picked up by the live microphones.

hard light A small, intensive light, producing strong highlights and dark shadows. See also *soft light.*

hardware The physical equipment.

haze filter A camera lens filter to reduce atmospheric haze by filtering out some ultraviolet and blue light. See also *contrast filter* and *correction filter.*

HBO Home Box Office cable television network (USA).

HDTV High-definition television.

head (1) The start, beginning of a tape or film. (2) Magnetic head; a tape recorder transducer. (3) Sound and picture head in editing machines. (4) Generic term for camera, printer, or projector mechanism without the accessories. (5) Tripod or camera head for mounting and/or moving the camera.

head alignment The correct position of the tape recorder head and gap in relation to the magnetic tape. Relates to height, Azimuth (tipping left and right), and Zenith (tipping forward and backward).

headend Control center of a cable television system where the programs, originating from different sources, are assembled or routed electronically.

headphones A pair of earphones worn by the soundmen/women and production personnel to monitor sound pick-up and/or to receive directions.

head room (1) The space between the top of the head of the performer or subject and the top of the picture frame. (2) Recording level; intentionally recording the picture or sound signals lower than normal to leave enough "head room" for sporadic loud or overly bright signals and to avoid clipping those short-duration signals, causing obvious distortion or compression.

headset Also called **cans.** A pair of headphones equipped with a microphone for communication. It is worn by the television camera operator, sound mixer, boom operator, floor manager and other crew members.

head shot   A tight close-up (CU) of the performer's head, filling the frame.

head slate   Clapboard recorded/filmed at the beginning of a take (shot). It may be an old-style clapboard or electronically generated image and sound. See also *end slate*.

helical scan   Also called **slant track.** Scanning pattern in a type of videotape recorder where the tape travels helically around a fixed larger drum, while a smaller drum carries a recording/playback head within. The helical wind causes a slant in the recording pattern. See also *quadruplex recording* and *transverse scanning*.

helicopter mount   Specifically designed camera mount used in helicopter (aerial) photography that counterbalances movements and vibrations. See also *Dyna lens*.

hem   Hemispherical beam.

henry (H)   Also called **the henry.** A unit of inductance in which an induced electromotive force of one volt is produced when the current is varied at the rate of one ampère per second. Named after Joseph Henry (1797-1878), American physicist.

hertz (Hz)   International unit of frequency; one hertz equals one cycle per second (cps). Named after Heinrich Rudolf Hertz (1857-1894), German physicist. See also *kilohertz, megahertz, Ghz*.

HF   High frequency.

Hi 8   Upgraded 8mm video tape format. See also *8mm*.

high-angle shot   Also called **top shot** or **bird's eye view.** Subject photographed from above, from an elevated point.

high band   Color videotape recording utilizing high carrier frequency that reduces electronic interference. Applies to quadruplex and helical scan videotape formats. See also *low band*.

high definition television (HDTV)   A format that makes large-size television pictures possible, in a 16:9 ratio. One popular format is the 1,125-line/60-field per second production system developed by Sony/NHK. Other formats offer 1,250, 1,050, or 895 scan lines. See also Appendix C's table of Existing Television Picture Standards, page 254.

high fidelity (hi-fi)   Accurate reproduction of music in a broad frequency range, up to 15,000 or 20,000 Hz, with no distortion and bass response as low as 30 Hz.

high frequency (HF)   3–30 MHz with a wavelength of 100 m. See also *low frequency*.

high hat (hi hat)   Also called **top hat.** Small camera mount, fixed either on a tripod head or placed on the floor on any low platform for low-angle shots.

high key   High intensity lighting; bright overall illumination on the set. See also *low key*.

highlight   Brightly lit subject or area.

highly directional mike   See *shotgun mike*.

high-speed camera   Film camera specially constructed to achieve slow-motion projection effect. It is capable of recording an image at a rate much faster than a conventional camera (up to one million frames per second and higher).

high-speed cinematography (HSC)   Specific type of cinematography using high-speed cameras for slow-motion effects, used in sports, science technology, and space to study and analyze fast moving events and transient phenomena that can not be recorded by other methods.

high-speed film   (1) Film with fast emulsion and high sensitivity to light. (2) Film manufactured specifically for high-speed cinematography.

"hit the mark"   Command to and move by an actor/performer to take up a predetermined spot (mark) on the set.

HKG   ITU country code for Hong Kong.

HMI   Halogen metal iodide; widely used enclosed-arc photographic lamp (5600K or 3200K), a trademark of Osram, now in common usage. The lamp source is extremely rich in ultraviolet energy.

HN   Headline News cable television network (USA).

HND   ITU country code for Honduras.

HNG   ITU country code for Hungary.

HOL   ITU country code for The Netherlands.

hold frame   See *freeze frame*.

"hold to BG/hold under"   See *background*/1.

holography   Pictorial image recorded on an emulsion surface provided by the coherent light of laser. The picture then can be viewed in three dimensions after a suitable reconstruction process. Holography was invented in 1947 by Dennis Gabor (1900–1979), Hungarian-born British physicist, who in 1971 was awarded the Nobel Prize for his discovery.

HONDUTEL Empresa Hondurena de Tele-comunicaciones (Honduras).

honey wagon Colloquial for a mobile dressing room, toilet, or trailer.

hot camera Activated, turned on camera. See also *dead*/1 and *live*/2.

hot mike Activated, functioning microphone. See also *dead*/1 and *live*/2.

hot spot Also called **womp.** An extremely bright, excessively lit, often reflective area of the scene, subject, or screen.

house light Regular (house) illumination in the studio.

housing See *underwater housing.*

Jack R. Howard Broadcast/Cable News Awards Awarded annually to single radio or television stations or a commonly owned group of radio stations for design and promotion of the public good by the Scripps Howard Foundation (USA).

hr Hour.

HR Hessische Rundfunk (Germany). See *ARD.*

HRV ITU country code for Croatia.

Hrvatska Izvjestajna Novinska Agencija Croatian News and Press Agency.

HS (1) High speed. (2) Hindustan Samachar (India).

HSC High speed cinematography.

HSN Home Shopping Network cable television network (USA).

HSS High-speed system.

HTI ITU country code for Haiti.

HTV Hrvatska Televizija (Croatia).

hue The dimension and a specific variety of color. Often called **tint** on TV receivers.

Hugh Malcolm Award See *National Association of Broadcasters.*

Hugo Award Annual award in 37 categories. See *Chicago International Film Festival.*

hum A distortion in sound or in visual image caused by the unwanted introduction of 50Hz or 60Hz power line energy into the desired signal.

hum filter   See *rumble filter.*

HUT   Households using television.

hybrid satellite   Satellite utilizing C-band and Ku-band transponders.

hyphenates   Colloquial for double or triple duties, usually hyphenated, carried out by a single professional, i.e., writer-producer, writer-director, director-cinematographer, producer-director, writer-producer-director.

hypo   Synonym for fixing bath in film development.

HWA   ITU country code for Hawaii.

I

I (1) ITU country code for Italy. (2) Information, informational.

IAB (AIR) International Association of Broadcasting (Uruguay).

IAFA International Animation Film Association.

IANA Inter-African News Agency (Zimbabwe).

I.A.T.S.E. Also called **IA**. International Alliance of Theatrical and Stage Employees and Moving Picture Machine Operators of the United States and Canada.

IBA (1) The former Independent Broadcasting Authority (GB). See *ITC*. (2) Israel Broadcasting Authority.

Ibero-American Television Organization Organizacion de la Television Ibero Americana—OTI (Mexico).

IBS International Broadcasting Society (Korea).

IC Integrated circuit.

ICA The one-time International Communication Agency, now called the United States Information Agency (USIA).

ICARTA (ACIRTA) International Catholic Association for Radio, Television and Audiovisuals. See *UNDA* (Belgium).

ICF Institut Canadien du Film.

ICIA International Communication Industries Association.

ICO ITU country code for Cocos-Keeling Islands.

iconoscope (Ike) Obsolete cathode ray-style television camera tube that made the first all-electronic television transmission possible. It was invented by Vladimir K. Zworykin (1889–1982), a Russian-born American physicist and electrical engineer. See also *Image Orthicon, Plumbicon, Vidicon* and *Saticon*.

ICRT Instituto Cubano de Radio y Television (Cuba).

ID Identification; station identification.

IDFA   International Documentary Film Festival-Amsterdam; Stichting Internationaal Documentair Filmfestival Amsterdam (The Netherlands).

idiot card   Idiot sheet. See *cue card.*

IDL   International Date Line.

ID signal   Identification signal; sound and visual signal placed on major international broadcasts to assist line-up. The visual signal carries the name of the source superimposed over the test card, while the sound identifies the source (in the original language) on a continuous loop.

IDTV   Improved definition television.

IEEE   Institute of Electrical and Electronics Engineers. (GB)

IETEL   Instituto Ecuatoriano de Telecomunicaciones. (Ecuador)

IF   Intermediate frequency.

IFJ   International Federation of Journalists.

IFPA   See *AVC.*

IFRB   International Frequency Registration Board (ITU).

IIC   International Institute of Communications (GB).

IINA   Ireland International News Agency.

IIP   International Informatics Programme of UNESCO. See *United Nations Educational, Scientific and Cultural Organization.*

Ike   (1) Colloquial abbreviation for iconoscope. (2) Short name for the Ikegami video camera.

illumination   See *lighting.*

illusion of movement   See *persistence of vision.*

IMA   Interactive Media Association (USA).

image   The photographic likeness of an object as it appears on the television picture tube, motion picture film, or on screen.

image area   The total surface covered by the image in the frame or on the screen.

image degradation   General loss of quality, contrast, and detail of a photographic image due to age, duplication, incorrect handling, or impurities.

image distortion   See *distortion*/2 and 3.

image intensifier   An adapter, an electrical device, placed between the camera and the lens to improve photography in low-key lighting.

Image Orthicon (IO)   A very sensitive monochrome television camera pick-up tube. Now obsolete. Nicknamed *Immy*. See also *iconoscope, Plumbicon, Vidicon* and *Saticon*.

image processing   Computer animation where actual images are processed. See also *synthetic processing*.

image retention   See *comet tail*.

image structure   Capacity measurement of an emulsion's detail imaging quality.

IMEVISION   Instituto Mexicano de Television; Mexican Institute of Television.

Immy   Popular nickname for *Image Orthicon*.

impedance   The changing resistance of alternating current (AC) frequencies measured in ohms.

improved definition television (IDTV)   The use of a frame store to interpolate extra lines from an NTSC signal. See also *high-definition television*.

IMTV   Interactive Multimedia Television.

IN   (1) Abbreviation for inch. (2) In-cue; the opening/starting words of a story or program. See also *out*/1. (3) In point; indication for an edit in video. See also *out*/2.

INA   (1) Iraqi National Agency. (2) Israel News Agency.

incandescent light   Bright (hot) light produced by glowing filaments in gas-filled lamps.

inch   British linear measure, equals 25.4mm or 2.54cm.

inching knob   A knob on film cameras, projectors, and editing equipment to move the film at short stops back and forth to check synchronization or proper loop alignment.

incidental sound   See *sound effects*.

incident light   Illumination falling upon a subject from external light sources.

incident light meter   Exposure meter that measures the light falling upon the subject. See also *reflected light meter*.

IND   ITU country code for India.

independent broadcast station   A commercially operated broadcast station with no network affiliation.

Independent Television Commission (ITC)   The British organization to license and regulate commercial television services from and to Great Britain, including satellite and public teletext. It replaced the Independent Broadcasting Authority and the Cable Authority (GB).

index beam   See *pilot beam*.

inductance   A conducting coil, a circuit element, in which electromagnetic force is generated by electromagnetic induction.

industrial film   Informational, usually a documentary type short film, showing certain aspects of industrial, manufacturing, or training operations.

industrial grade   Related to the quality of a camera, monitors, and video recorder systems. See also *broadcast quality* and *consumer grade*.

infinity   Lens distance calibration scale showing the distance beyond which light rays appear to be parallel.

infomercials   Information-commercials; program-length broadcast or cablecast advertisements for various products.

informational program   Primary television news program or film of various types and length of factual and current events.

infotainment   A somewhat artificial name in vogue for information and entertainment broadcasts. It refers to any information program with entertainment as its underlying theme.

infrared   Invisible rays of the visible spectrum beyond red with longer wavelengths and with penetrating heating effect.

infrared cinematography   Filming on infrared sensitive film material using a corrective lens in the dark or in day-for-night cinematography. See also *night effect*.

inherent noise   The sound (noise) of the projection or display system.

in-house propagation   See *carrier current*.

inky or **inkie**   Colloquial term for a small incandescent light.

inky dinky   Also called **dinky inky**. Small, incandescent studio lamp with low wattage.

inlay   Part of one television camera picture inserted into another. See also *insertion* and *overlay*.

in-line heads Also called **stacked heads.** Stereophonic heads arranged in a tape recorder so that the head gaps were mounted directly above each other. Now obsolete.

INMARSAT A consortium of member nations (67) operating a global satellite system in maritime, aeronautical, and land mobile communications (GB).

INN Independent Network News (GB).

input Camera, mixer, or recorder terminal receiving signals or electric power. See also *output*.

INRAVISION Instituto Nacional de Radio y Television (Colombia).

INRT (INRTV) Iranian National Radio and Television.

INS (1) ITU country code for Indonesia. (2) International News Service.

insert (1) Insert shot; a close-up, stationary shot of an explanatory subject like a book, letter, or photograph. (2) Insert edit; a section of the finished program where audio and video have been edited into a segment.

insertion A special effect in television. See *inlay* and *overlay*.

instant replay Key action of a remote (OB) broadcast, like a sports event, recorded on a special videotape recorder for immediate (instant) playback, often in slow motion.

Institut Canadien du Film (ICF) See *Canadian Film Institute.*

Institute of Radio Engineers (IRE) A professional organization that establishes standards for radio and television engineers and publishes relevant informational material (USA). See also *Society of Motion Picture and Television Engineers.*

Instituto de la Cinematografia y las Artes Audiovisuales Institute of Cinematography and Audiovisual Arts, of the Ministry of Culture (Spain).

instructional television or **program** See *educational broadcasting/film.*

Instructional Television Fixed Service (ITFS) Special FCC-approved radio spectrum space reserved for US-domestic short-distance, low-power broadcast television signals for instructional purposes. Located in the 2400 MHz to 2700 MHz band.
See also *multipoint distribution service* and *multiple-multipoint distribution service.*

in-sync Properly aligned sound and picture. See also *off-sync.*

INT Interior.

integral tri-pack   See *tri-pack.*

INTEL   Instituto Nacional de Telecomunicaciones (Panama).

INTELCAM   Societé des Télécommunications Internationales du Cameroun.

Intelsat   International Telecommunications Satellite Organization (USA).

intensity   (1) High degree of light illumination, measured in candelas. (2) Increased opacity of the deposit on a negative. (3) High volume (loudness) of sound.

interactive television (ITV)   Two-way communication through cable television that also allows the viewer to select or access several different video or audio programs.

intercom   Intercommunication; two-way communication between the control room and the studio personnel.

intercut   A single shot or two independent sequences related in a meaningful way.

interdupe   See *intermediate.*

interference   Disturbance of radio and/or television reception caused by undesirable external sources.

interior (INT)   Scene, shoot, or program taking place indoors. See also *exterior.*

interlace   Interlace scanning; the interweaving of two sets of parallel lines to form a complete picture and to reduce flicker in the television receiver.

interlock   A system to ensure synchronous projection of separate video picture or film (the edited workprint) and the matching sound track. See also *Selsyn motor.*

intermediate   Also called **interpositive** or **interdupe** (if negative). Intermediate film; a positive or negative film from an integral tri-pack color negative.

intermittent movement   The motion of the film in the camera or projector in a stop-and-go fashion, stopping each frame at the gate to be exposed or viewed as a still frame and then replaced by the following frame.

internal battery   See *battery.*

International Affiliation of Writers Guilds   An affiliation of national writers' organizations (guilds) that includes the Writers Guild of

America-East, Writers Guild of America-West, Writers Guild of English- and French-speaking Canada, and the Writers Guild of Great Britain, Australia and New Zealand.

International Documentary Festival-Sheffield   International festival celebrating the themes of "Cinéma vérité," "USA" and "Under Fire" (film crew and directors filming in combat conditions), held in Sheffield, (GB).

International Documentary Film Festival-Amsterdam (IDFA)   Stichting Internationaal Documentair Filmfestival—Amsterdam. Annual festival for short, television, and documentary films, presenting the Joris Ivens Award. (The Netherlands).

International Documentary Film Festival Arsenal   International film festival for documentary films held in Riga (Latvia).

International Festival of Media Arts   See *MEDIAWAVE.*

International Film Festival—Karlovy Vary   Annual documentary and information film festival (Czech Republic).

International Film Festival—Locarno   International competition for premieres of feature and documentary films (Switzerland).

International Film Festival—Strasbourg   Competitive film festival for fiction and documentaries featuring human rights topics, held in Strasbourg (France).

International Filmfestspiele—Berlin   Competitive international film festival, presenting the Berlin Bear Prizes, held in various categories. The festival also features the European Film Market (Germany).

international news   Newscast consisting of news stories from abroad, filed usually by international news agencies, or by network or station correspondents. See also *local* and *national news.*

International Odense Film Festival   Children's, experimental, and fairy tale festival for films less than one hour long, held in Odense, Denmark.

International Radio and Television Organization (IRTO)   Organisation Internationale du Radio-Télévision-OIRT; an organization of the former East European countries, now merged with the EBU. See *European Broadcasting Union.*

International Science and Technology Film and Video Festival   Competitive festival highlighting science and technology, education, nature, and environmental topics held in Tokyo (Japan)

International Short Film Festival Oberhausen International competition of short films in fiction, documentary, animation, and experimental categories held in Oberhausen (Germany).

International Sports Film Festival—Budapest International film festival for sports films held each fall in Budapest (Hungary).

International Standardization Organization (ISO) The coordinating organization of several national groups that sets and establishes standards and issues relevant publications for the motion picture and television industry. It is based in Washington, DC (USA). See also *ASA*.

International Telecommunications Satellite Organization (Intelsat) Founded in 1964 and based in Washington, DC, the organization began with the launch of Early Bird, a small communication satellite that initiated satellite service for the North Atlantic. Today, with coverage over the Atlantic, Pacific and Indian oceans, some 2,200 separate communication pathways provide service for telephone, television, facsimile, and data communication to nearly 180 countries, dependencies, and territories: See also *COMSAT*. Current and future Intelsat space crafts include:

INTELSAT V (1980)
INTELSAT V-A (1985)
INTELSAT VI (1989)
INTELSAT K (1992)
INTELSAT VII (1993)
INTELSAT VII-0A (1995)
INTELSAT VIII (1996)

International Telecommunication Union (ITU) Unité International des Télécommunications-UIT, (Geneva, Switzerland); founded in 1865 as an intergovernmental organization, ITU became a specialized agency of the United Nations in 1947 and has 172 member countries. Its responsibilities include planning and regulating worldwide telecommunications, equipment standards, systems operating standards, and the dissemination of pertinent information.

International Television Festival of Monte Carlo Presenting the Golden and Silver Nymphs, the international event is held annually in Monte Carlo for television programs and animation (Monaco).

International Writers Guild See *International Affiliation of Writers Guilds*.

internegative Intermittent negative; color duplicate negative film produced directly from the reversal color master original. Used for making ("running") release prints.

interpositive   A duplicate positive film used for making additional prints.

Intersputnik   International satellite organization founded in Eastern Europe in 1971 with 16 member nations for communications operation and inter-country video exchanges (Moscow, Russia.)

intersync   Accessory to older quadruplex videotape recorders for smooth integration of mixing between two or more videotaped sources.

intervalometer   Master control that operates camera, lights, and auxiliary shutter at predetermined intervals. Used in animation timelapse cinematography.

interview program   Unrehearsed broadcast talk program aimed to explore the character or events around a person or specific knowledge of an expert guest. Interviews may be part of an informational (news) program or may be scheduled as a separate show.

intervision   Phase reverse. See *negative projection.*

"In the can"   Expression denoting a completed broadcast program or film.

INTV   Association of Independent Television Stations (USA).

inverter   A special device that changes the type of electrical current (typically 12 VDC or 24 VDC currents from autos, trucks, boats, and aircraft) to appropriate AC current at proper voltage.

IO   Abbreviation for Image Orthicon.

IOB   ITU country code for St. Kitts & Nevis, St. Lucia, St. Vincent & the Grenadines, Turks & Caicos Islands and the British Virgin Islands.

ionosphere   The region of the earth's upper atmosphere at an altitude of 30–90 miles (appr. 50–150 km.) and above, containing electrically charged particles (ultraviolet radiation, X-rays from the sun) that plays an important role in intercontinental radio transmission.

ionospheric wave   See *sky wave.*

IOR   Indian Ocean Region; satellite ocean region configuration.

IPDC   International Programme for the Development of Communication of UNESCO. See *United Nations Educational, Scientific and Cultural Organization.*

IPMPTI   International Photographers of the Motion Picture and Television Industries (USA). See *I.A.T.S.E.*

ips   Inches per second.

IRCC   International Radio Consultative Committee; Comité Consultatif Internationale des Radio Communications—CCIR (Switzerland).

IRE   Institute of Radio Engineers (USA).

IRIB   Islamic Republic of Iran Broadcasting.

iris   Mechanism for adjustable lens opening, named after the contractible membrane of the human eye. See *diaphragm*.

iris wipe   Also called **iris in** and **iris out**. A gradual appearance or disappearance of an image from a small spot or into a small spot in the form of a circle (circle in or circle out). In television it is achieved electronically, in film by a laboratory process.

IRL   ITU country code for Ireland.

IRN   (1) ITU country code for Iran. (2) Independent Radio News (GB).

IRNA   Islamic Republic News Agency (Iran).

iron oxide   Also called **oxide.** A microscopic ferrous oxygen compound used for coating magnetic tape in manufacturing.

IRQ   ITU country code for Iraq.

IRTC   The Independent Radio and Television Commission (Ireland).

IRTO   The former International Radio and Television Organization; Organisation Internationale de Radio et Television-OIRT. Now EBU-European Broadcasting Union.

IRTS   International Radio and Television Society (USA).

ISBO   Islamic States Broadcasting Services Organization (Saudi Arabia).

ISL   ITU country code for Iceland.

island   See *film island.*

ISO   International Standardization Organization.

ISR   ITU country code for Israel.

ITA   Independent Television Authority, a former organization in Great Britain. See *IBA.*

ITAR-TASS   International Telegraphic Agency of Russia. See *TASS.*

ITC   Independent Television Commission (GB).

ITFS   Instructional Television Fixed Service.

ITIM   News Agency of the Associated Israel Press.

ITN   (1) Independent Television Network (Sri Lanka). (2) Independent Television News Service (GB).

"It's a print"   Film director's command to process and to workprint a successful take. Also known as **"Print it."**

"It's a wrap"   See *wrap.*

"It's on the air"   See *"Take it away."*

ITU   International Telecommunication Union; Unité International des Télécommunications-UIT.

ITU country code   Abbreviated code for the countries of the world allocated by the International Telecommunication Union.

ITU region   The three world regions as devised by the International Telecommunication Union: Region I: Europe, Africa, Near and Middle East; Region II: the Americas; Region III: Asia, the Indian subcontinent and the South Pacific.

ITV   (1) Instructional television. (2) Interactive television. (3) Independent television.

I.U.E.   International Union of Electrical Workers (USA).

IVCA   International Visual Communications Association (GB).

IVCA Film and Video Communications Festival   Yearly festival for non-broadcast productions held by the International Visual Communications Association in London (GB).

IVDS   Interactive video data service (USA).

Joris Ivens Award   Annual documentary film awards presentation. See *International Documentary Film Festival-Amsterdam* (The Netherlands).

IWG   International Writers Guild. See *International Affiliation of Writers Guilds.*

J   ITU country code for Japan.

jack   Also called **black jack.** (1) Socket for electrical, telephone, or headset connection. (2) Stage brace. (3) Actuator that moves the satellite dish eastward or westward.

jam   Also called **spaghetti.** Jammed film in the camera. See *buckle.*

JAMINTEL   Jamaica International Telecommunications, Ltd.

jamming   The intentional interference of a signal from another station or transmittal source by unusual means.

JASRAC   Japanese Society for Rights of Authors, Composers, and Publishers.

JBC   (1) Jamaica Broadcasting Corporation. (2) Japan Broadcasting Corporation. See *NHK.*

JCET   Joint Council on Educational Television (USA).

JDC   Japanese digital cellular.

jeeped monitor   Old-style wheeled studio monitor.

jelly   Colloquial for (1) Color correction filter. See *gelatin filter.* (2) Light diffuser.

jenny   Colloquial for electrical generator.

jet   See *air knife* or *squeegee.*

jibbing   See *tongue.*

jingle   A repetitious, lively musical piece, mostly song, used as a musical theme or for commercials.

JMC   ITU country code for Jamaica.

Jockey roller   A roller in the film projection system that keeps the film under required tension.

jogging   Videotape frame-by-frame movement in slant track (helical scan) editing.

JOR   ITU country code for Jordan.

joule   International system unit of energy; equals the current of 1 ampere passed through 1 ohm of resistance for 1 second. Named after James Prescott Joule (1818–1889), British physicist.

JP   Jiji Press (Japan).

JRT   Jugoslovenska Radiotelevizija (Yugoslavia).

JRTV   Jordan Radio & Television Corporation.

JSA   Japanese Standards Association.

JSB   Japan Satellite Broadcasting Corporation.

JTA   Jewish Telegraphic Agency (Israel).

juice   Slang for electrical power.

juicer   Colloquial for electrician.

jump cut   A cut (splice) in the tape or film interrupting the normal flow of continuous action. It may be unintentional or done for effect.

jumping cue   A performer starting action or speech earlier than his/her cue.

junior   A 2,000 watt studio spotlight. See also *senior*.

K (1) Kelvin. (2) First letter of broadcast station call letters west of the Mississippi, and KDKA in Pittsburgh, Pennsylvania, and KYW in Philadelphia, Pennsylvania (USA). (3) Memory chip storage capacity representing approximately 1,000 memory cells (1,024 actual).

Ka-band A very high frequency band of 18–20 GHz used in satellite communications enabling satellite antennae to form narrower beams and cast smaller shadows on the earth (150 mi. or 241.3km in diameter versus the size of the Continent below).

KAZ ITU country code for Kazakhstan.

KBS Korean Broadcasting Corporation (Republic of Korea).

Kc Kilocycle.

KCN Keskustapuolueen Sanomakeskus (Finland).

KDKA The first commercially licensed radio station in Pittsburgh, Pennsylvania (USA). KDKA was founded in November 1920 by Dr. Frank Conrad, a research engineer, and H. P. Davis, a Westinghouse Corporation vice president. See also *8XK* and *experimental radio*.

KEG A small spotlight. See *baby*.

Kelvin, Kelvin scale (K) Unit of thermodynamic temperature. Temperature scale in Celsius, starting with absolute zero (approximately –273.16° C). Named after Lord William Kelvin (1824–1907), British mathematician, physicist, and inventor. See *color temperature*.

KEN ITU country code for Kenya.

Byron Kennedy Award Yearly award by the Australian Film Institute given to encourage unorthodox and visionary approach to film making. See *Australian Film Institute*.

key (1) Special effects signal used in electronic matting. (2) Intensity of a light source. See *high key* and *low key*. (3) Denoting, in general, the head, the leader of a group or unit.

key announcer   See *anchor* or *linkman/woman*.

key grip   Head grip.

key light   Also called **modeling light.** Primary source of illumination. See also *fill light, back light* and *background light*.

key numbers   See *edge numbers*.

kg   Kilogram.

kg/cm²   Kilogram per square centimeter.

KGZ   ITU country code for Kyrgyzstan.

KHS   Kaigai Hyoron Sho (Japan).

kHz   Kilohertz.

kicker light   Light directed from the side and the back of the subject.

kill   Colloquial term. (1) To eliminate or cancel a project or program. (2) To remove or strike certain parts of the scene or the set. (3) To shut off lights and/or equipment.

kilocycle (kc)   1,000 oscillations per second. See *kilohertz*.

kilogram (kg)   1,000 grams, the international unit of mass. Equals appr. 2.2 pounds.

kilohertz (kHz)   1,000 hertz.

kilometer (km)   1,000 meters, equals 1,094 yards or 0.6214 mile.

kilowatt (kW)   1,000 watts.

kine   Abbreviation for kinescope (cathode ray) tube.

Kinemacolor   The first successful cinematographic color process (1906), now only of historical value.

Kinematographe   A last-century projector designed in 1896 by Oskar Messter (1866–1943), German inventor.

kinematoscope   An early device with still pictures placed on a wheel and rotated rapidly to give the illusion of motion, designed and patented by Coleman Sellers (1827–1903), American engineer.

kinescope recording   Also called **kine.** Film recording on motion picture film stock directly from a television picture tube. Replaced by videotape recording.

Kinetoscope   A peep-show viewer developed around 1893 by Thomas Alva Edison (1847–1931), American inventor, for 35mm film used in a camera devised by W. K. L. Dickson in 1891.

KIPA Katolische Internationale Presse Agentur (Switzerland).

KIR ITU country code for Kiribati.

Kliegl light Trade name for a powerful studio arc light, named after the inventor brothers John H. (1869–1959) and Anton T. Kliegl (1872–1927), American lighting experts.

km Kilometer.

km/s Kilometer per second.

KNA Kenya News Agency.

knee shot A medium type shot, showing the person(s) from the knee up.

KNR Kalaallit Nunaata Radioa (Kalaallit Nunaat).

Kodachrome 16mm consumer (amateur) type integral tri-pack color positive film, made in the United States of America. Received extensive usage in the Super 8 format.

Kodacolor A negative/positive color film process used solely in still photography (USA).

Kodak standard (KS) See *positive perforation*.

Kopernikus German satellite system.

KOR ITU country code for Korea (South).

Alexander Korda Award Annual presentation for the Outstanding British Film of the Year during The Lloyds Bank BAFTA Performance Awards. See also *British Academy of Film and Television Arts*.

KRE ITU country code for Korea (North).

KS (KS Standard) Kodak standard. Also called positive perforation. See also *BH*.

KSA Key second assistant.

KTN Kenya Television Network.

Ku-band Frequency band of values 10.7–12.75 GHz for Satellite Communication and Direct Satellite Broadcasting.

> Ku1-Band 10.7–11.75 GHz
> Ku2-Band 11.75–12.5 GHz
> Ku3-Band 12.5–12.75 GHz (Telecom)

kukie Kookie; see *cucalorus*.

kW Kilowatt.

KW Kurtzwelle; shortwave. Abbreviation found on German-made radio receivers.

KWT ITU country code for Kuwait.

Kyoto The Kyoto News Service (Japan).

L Lambert

lab Laboratory.

laboratory Lab; the place where motion picture processing, developing, grading, printing, and related services are carried out.

laboratory film Intermediate film used in various stages in the laboratory. See also *camera film* and *print film*.

lag A smearing of the television picture while there is motion in the scene. Typical of Vidicon, Plumbicon, and Trinitron image tubes.

Lambert (L) Unit of brightness in the centimeter-gram-second (CGS) system; equals one lumen per square centimeter. Named after Johann Heinrich Lambert (1728–1777), German physicist, mathematician, and astronomer.

lamp Light; often refers to a light bulb.

landline A special cable or telephone line used in remote broadcasts.

lanterna magica See *magic lantern*.

lanyard microphone See *lavalier*.

LAO ITU country code for Laos.

lap dissolve See *dissolve*.

lapel Lapel microphone; a small microphone worn around the neck or on the lapel. See *lavalier (lav)*.

lap splice Lap dissolve; the splicing together of two pieces of film in a way that they overlap. Splices may be of varying widths—negative or sound splices being narrower and positive splices wider. See also *butt splice*.

large-screen television Large screen projection; television projection forming large images by either

1. Cathode ray tube projection
2. Eidophor

3. Liquid crystal display
4. Plasma display.

Larsen effect   Acoustic feedback, a "howling" sound as the signal passes from the loudspeaker to the microphone, resulting from incorrect microphone–speaker placement.

laser   Light amplification by stimulated emission of radiation; an optical maser. Visible or near-visible beam of the spectrum used in wide-band communication.

latent image   Invisible (undeveloped) image on the exposed film emulsion that becomes visible in the development process.

latitude   The scope of exposure of film emulsion that renders a satisfactory image reproduction.

lavalier (lav)   A small microphone usually worn on the lapel or on a holding string around the neck by the announcer or performer. See *lapel*.

lavender net   Least dense, open-end net on a light. See also *open-end net*.

lazy arm   Lightweight microphone support with a pantograph-like arm.

lazy boy   See *pantograph*.

LBN   ITU country code for Lebanon.

LBR   ITU country code for Liberia.

LBS   Liberian Broadcasting System.

LBY   ITU country code for Libya.

LCA   ITU country code for St. Lucia.

LCD   Liquid crystal display.

lead   The leading (principal) character or the principal role in a teleplay or film.

leader   (1) Blank tape and/or film at the beginning of magnetic tape or film attached for cueing purposes. On film it may be black, opaque, or numbered. See *Academy leader* and *videotape leader*. (2) Black film used for spacing in the workprint. (3) Machine leader; strong blank film used to thread film in developing machines.

David Lean Award   A yearly Lloyds Bank BAFTA Performance Awards presentation for the best achievement in direction (GB). See also *British Academy of Film and Television Arts*.

LED   Light-emitting diode(s).

legal release   A legal agreement that conveys the right to use the name, likeness, property, and testimonial of a person for broadcast, film, and/or publicity purposes. In case of a minor, the release must be signed by the parent or legal guardian. See also *model release* and *property release*.

legs   See *tripod* and *baby legs*.

lens   A series of finely-ground, curved precision glass elements in a housing that concentrate light rays to form an image on the image sensor or film.

lens adapter   A camera mount that enables lens interchanges and ensures proper seating of the lens.

lens aperture   See *aperture*.

lens barrel   A tubular lens housing that includes a diaphragm, f-stop (T-stop) indications, and a focusing ring. Zoom lens barrels also carry a manual or electronic zooming device.

lens cap   Also called **cap.** Lens cover of metal, rubber, or plastic material.

lens coating   See *coating/1*.

lens cradle   A lightweight support for long focal length lenses.

lens hood   A cylinder-shaped shield placed in front of the camera to prevent undesired light rays from entering the lens.

lens opening   See *f-number*.

lens shade   See *eyebrow* and *sunshade*.

lens speed   See *f-stop*, i.e. f-number.

lens turret   Round-shaped plate mounted in front of the camera to hold a zoom lens. In old style, multiple-lens cameras, three lenses (in film cameras), and four lenses (in television cameras) rotated into position. Multiple lens turrets are being replaced by zoom lenses.

LES   Land earth station.

level   (1) Volume of sound. (2) B&W picture intensity in video. Related to the volume and gain of the desired signal to be recorded.

level indicator   See *magic eye, LED,* and *VU meter*.

lever   A straight handle fixed on the sleeve of a zoom lens to facilitate smooth movement.

LF   Low frequency (long wave).

library footage   See *stock footage*.

library music   Music stored and cataloged in a music library. See *music library*.

library shot   A shot or film sequence in or from the videotape/film library. See *stock footage*.

libretto   Book or lyrics for a large musical work.

license   (1) Station license. See *FCC*. (2) Musical license. See *ASCAP* and *BMI*. See also *copyright, model release* and *property release*.

light balancing filter   A filter used to attain true, adequate color balance.

light box   A frosted-glass-covered illuminating box used at the editing table for viewing film footage and/or slides.

light diffusing screen   See *butterfly, gauze,* and *scrim*.

lighting   The illumination of a scene or subject using the photographic principles of key light, fill light, and back light in both television and motion picture production. See also *back light, fill light* and *key light*.

lighting cameraman/woman   Cinematographer or director of photography who also handles lighting. British term.

lighting contrast ratio   The ratio between the key light and the fill light.

lighting grid   See *grid*.

lighting plan   A floor plan-based layout indicating the type and position of lights and their beam direction.

light level   The overall intensity of illumination measured in candela.

light lock   A chamber configuration with double doors, curtains, and baffles that protect the entrance of a darkroom from light.

light meter   See *exposure meter*.

light speed   See *velocity of light*.

light trap   Device preventing light from falling on the film in cameras and printers. See also *baffle/2*.

light units   Units by which the intensity of the quality of light are measured. See *angstrom*.

liko   See *ellipsoidal spotlight*.

lily   Color pattern photographed onto either end of the film reel as a guide for color standards in the laboratory.

limbo   Any background with no visual significance (usually gray or black).

line   (1) Telephone line used for program transmission. (2) A written and/or spoken line of a script or announcement.

line input   See *input*.

line microphone   Directional microphone.

line monitor   See *master monitor*.

line producer   Production executive, often with duties of a production manager.

line-of-sight transmission   See *point-to-point*.

line output   See *output*.

liner   See *station identification*.

line rate converter   Converter employing a seven-head recorder designed to convert line rates, e.g. 625-line to/from 525-line. Used for color television pictures.

line store converter   Converter that writes the incoming information into suitable storage, then reads out the video signal at the rate required by the output scanning standard.

line-up time   Also called **warm-up time.** (1) Time period required for warming up and adjusting the equipment in a television studio. (2) Time required for the film camera crew to set up and check the camera equipment and prepare for shooting.

linkman/woman   Announcer connecting different broadcast materials in a continuous program. See also *anchor*.

lipstick camera   See *point-of-view camera*.

lip sync   Synchronization of sound with lip movements.

liquid crystal display (LCD)   Liquid "sheet" between layers of plastic or glass that becomes opaque when electrical current passes through it. LCD displays are used for digital clocks, display panels, etc.

liquid gate   Special technique employed in some film projection and printing at the gate in order to prevent or reduce abrasion or scratches. In a telecine a wet gate projection technique masks existing scratches. See also *wet gate printer*.

liquid head   See *fluid head*.

live   (1) Transmission of a studio or remote (OB) program simultaneously with the actual performance. (2) Camera or microphone that is turned on; hot. See also *open* and *dead mike*.

live action   Actual performance, happening, action, or play. It may be broadcast directly or recorded or filmed for transmission/projection at a later date. See also *animation*.

Livingston Awards, The   Annual awards given by the Mollie Parnis Livingston Foundation to young journalists (aged 34 and under) for broadcast coverage of national and international news. Administered by the University of Michigan (USA).

LJB   Libyan Jamahiriya Broadcasting.

Lloyds Bank BAFTA Performance Awards, The   Awards presentation in several categories in both film and television given jointly by Lloyds Bank and the British Academy of Film and Television Arts (GB). See *British Academy of Film and Television Arts*.

lm   Lumen.

LN   Local news.

LNA   (1) Low-noise amplifier. (2) Libyan News Agency.

LNBS   Lesotho National Broadcasting Service.

LNT   Low-noise tape.

LOC   Location.

local channel   A limited operational channel on which several lower-watt stations may operate.

local coop   Local cooperative advertising.

local news (LN)   Newscast comprised of news items from the area covered by the station as opposed to national or international news.

local program   Program originating at and transmitted by the station's own facilities.

local time (LT)   The hour-minute-second actual time of the area (locality) compared to other area time zones.

location   A place or site outside the television or film studio where production (shooting) is done.

location scouting   Pre-production activity of searching for and selecting suitable sites for recording and filming.

log   Program (pre-) log; detailed, second-by-second breakdown of a station's complete daily program schedule.

logo   Identification symbol of a broadcast station or company (advertiser).

log sheet   See *camera report.*

London Film Festival, The   A major non-competitive festival organized by the British Film Institute and held in its National Film Theatre in London (GB). See *British Film Institute.*

long distance (DX)   Radio and television signals received over long distances, i.e. shortwave radio.

long focal-length lens   Long lens. See *telephoto lens.* See also *short focal-length lens.*

Raymond Longford Award   Annual award presentation by the Australian Film Institute to an individual who has made a significant contribution to Australian film-making. See also *Australian Film Institute.*

long pitch   See *pitch*/2 and 2a.

long playing (LP)   Phonograph or gramophone record speed of 33½ rpm, usually on 12-inch disks.

long-range microphone   A highly directional dynamic microphone with a cardioid pick-up pattern capable of long-distance sound pick-up.

long shot (LS)   An establishing or cover shot that includes a large field of view. See also *wide angle shot.*

long wave (LW)   See *low frequency.*

loop   (1) A two-way connection (circuit) between the console and the broadcast location. (2) An endless, spliced, magnetic tape or film used continuously for dubbing of effects. (3) See *film loop.*

lot   A large land area, part of the studio complex, where exterior sets are constructed. The term is also used to indicate the entire studio area.

loud haler   British term for **bull horn.**

loudness control   Also called **contour.** The control that compensates for the loss of tones at the extreme end of the audio range when played at low volumes through small loudspeakers.

loudspeaker   Apparatus that converts electrical impulses into audible sound waves.

low-angle shot   A photographic shot taken from a low point, looking up on the subject. See also *high-angle shot.*

low band   Early quadruplex video tape recording. See also *high band*.

low-con   Low contrast.

low frequency (LF)   30–300 kHz with a 10 km wavelength. See also *high frequency*.

low key   Low intensity illumination emphasizing only the performer and/or creating a mood effect. See also *high key*.

low noise amplifier (LNA)   Device that magnifies the weak signals of the satellite receiving antenna.

low speed   See *fast motion*.

LP   (1) Leading personality. (2) Long playing phonograph record. (3) Long play speed on VHS video cassette recorders.

LRT   Lietuvos Radijas ir Televizija (Lithuania).

LS   Long shot.

LSO   ITU country code for Lesotho.

LT   Local time.

LTU   ITU country code for Lithuania.

lubrication (lub)   Lubrication applied on the processed film for best (maximum) projection life.

lumen (lm)   Unit of luminous flux. The basic quantity of light produced by one candle of one candela intensity at a unit distance.

luminaire   Lighting unit; lamp, housing, fixture, and/or accessories.

luminance   Luminous intensity; the reflected light of the visual image on the screen. Brightness expressed in candelas per square meter.

luminance channel   Part of the frequency spectrum of a color transmission or video recorder containing luminous information. Used for brightness control for both monochrome and color television receivers.

luminous flux   The time rate of light emission. The amount of light passing through an area in one second.

LUX   ITU country code for Luxembourg.

lux (lx)   Metric unit of illumination; one luminous flux of one lumen per square meter. See also *candela*.

LVA   ITU country code for Latvia.

LVR   Laser Videodisc Recorder.

LW Long wave.

LWF Lutheran World Federation Radio (Switzerland).

lx Lux.

lyricist Writer and/or poet who creates the words for musical compositions.

lyrics Poem or words to a song or musical number.

M

M   Medium wave, i.e. AM.

m   Meter.

MAC   (1) ITU country code for Macau. (2) Multiplexed analog component.

machine leader   See *leader*/3.

macro cinematography   ECU (extreme close-up) cinematography of small objects, still visible to the eye, that can be focused by the camera lens at close range without the use of a microscope.

macro lens   Specially designed lens used in macro cinematography capable of focusing at very close ranges, from a distance of appr. one inch (25mm) to infinity.

macrozoom   Camera zoom lens used in macro cinematography as opposed to a fixed focal length lens. A macrozoom lens facilitates finding a position; however, in macro mode the zooming cannot be used.

made-for-television features   Also called **tele-movies.** Films of feature length and subject made for television broadcast.

Madison Award   A yearly award given to a broadcast journalist or group of journalists for distinguished work in the protection and preservation of the freedom of expression. Given by the National Broadcast Editorial Association (USA).

magazine (mag)   Light-proof film container attached to cameras or projectors. Constructed of durable, lightweight material, it contains a section for raw stock (feed reel) and another for exposed or projected film footage (take-up reel).

magazine format   Television broadcast program presenting a variety of different features in a daily or weekly schedule.

magenta   Reddish-blue, green-absorbing (minus green) subtractive primary color used in three-color processes.

mag film   See *magnetic film*.

magic eye   A small bulb in a recording apparatus that indicates levels and serves as a warning light against under- or over-recording. Now obsolete.

magic lantern (lanterna magica)   An early projection device containing a lamp fixed on slides through a lens opening and used for projecting images.

magnetic film   Also called **mag film** or **sound stock.** Full coat magnetic tape (16mm and/or 35mm) with sprocket holes on the edges onto which sound, sound portion is transferred from 1/4 inch tape or digital audio tape (DAT) to facilitate synchronization in editing.

magnetic head   A transducer in magnetic recorders for the erasing, recording, and playback of electrical impulses, i.e. sound waves.

magnetic/optical projector   See *optical/magnetic projector.*

magnetic recording   Also called **tape.** A type of video and sound recording affecting magnetic variations in a coated (iron oxide) medium for recording magnetic impulses from audio and/or visual sources.

magnetic stock   Magnetic film.

magnetic stripe   A clear film with an iron oxide strip for sound recording and reproduction.

magnetic tape   Also called **audio tape** or **videotape.** Thin-base tape material of specified width with iron oxide coating used for sound and video recording.

magnetic tape recorder   See *tape recorder.*

magnetic tape speed   The rate of speed at which the tape travels through a recorder/playback machine. The standard speeds are:

$1\frac{7}{8}$ ips—4.76 mm
$3\frac{3}{4}$ ips—9.52 mm
$7\frac{1}{2}$ ips—19.05 mm
15 ips—38.01 mm
30 ips—76.02 mm

Professional tape speeds are:

$7\frac{1}{2}$ ips and 15 ips

magnetic tape width   The standard width of tape expressed in inches and/or millimeters:

$\frac{1}{16}$ in.—1.58 mm
$\frac{1}{8}$ in.—3.17 mm
$\frac{1}{4}$ in.—6.35 mm
$\frac{1}{2}$ in.—12.07 mm
$\frac{3}{4}$ in. —19.05 mm
1 in. —25.04 mm
2 in.—50.08 mm

magnetic track Also called **mag track.** Magnetic sound track; sound recorded on magnetic film or on magnetic sound strip along the side of the edge of the film base. See also *balance stripe.*

magniscale A scale model built to larger size than the original, allowing the television or film camera to show details effectively.

mag-opt Magnetic-optical print; a release print provided with both a magnetic and optical soundtrack for either type of projection system. See also *optical/magnetic projector.*

mag stripe See *magnetic stripe.*

mag track See *magnetic track.*

main title Caption and graphic material with the title and primary information of the television or film program. See *title.*

make-up (1) The transformation of the physical appearance (face and/or body) of the performers by the make-up artist into the characters to be portrayed. (2) The assembled, combined films on a large reel, usually prepared for a preview.

make-up artist Make-up man/woman; the artist who uses cosmetic material and means to create physical (facial/body) changes on the actors.

Maltese cross A slotted driving wheel in the shape of a Maltese cross, an essential part to maintain intermittent movement in a film projector.

MAP Maghreb Arabe Presse (Morocco).

MAR Metro area rating (metropolitan area rating).

Marconi Radio Award See *National Association of Broadcasters.*

mark (1) A piece of masking tape or color tape placed on the floor of the studio/location for exact placement of scenery, cameras, and/or actors, usually as a box mark, a "T", a toe mark, an upside down "V," or just a straight line. (2) Mark on the barrel of the camera lens effecting lens opening and change of focus, or zoom position in a continuous shot.

marketing   See *sales department.*

marking pencil   Also called **china marker, grease pencil,** or **wax pencil.** A soft pencil that produces a removable mark, used in editing and marking film.

married print   A positive film print with both the sound and picture. See also *composite print.*

maser   Microwave amplification by simulated emission of radiation; a low-noise microwave amplifier.

mask   See *Academy mask.*

masking tape   A crepe-finished adhesive paper-based tape in natural or other colors, in various widths and sizes for studio, graphic art, or editing room use.

mass communication   Communication used to reach, inform, and influence a large number of people through the mass media.

mass media   The various means of communication, i.e. the press, radio, television, and motion pictures, that reach a large audience.

mast   Antenna.

master   (1) An approved, completed script. (2) Official program schedule. (3) Original record or transcript die used for making duplications. (4) The original, completed sound or videotape recording. (5) The final negative, reversal positive, intermediate, or internegative film used to make subsequent prints.

master control   The centralized control where all studios, remote broadcasts, picture and sound sources are channelled for a complete and smooth presentation and transmission; the operations center often assisted by computer technology.

master monitor   Also called **actual monitor** or **line monitor.** The monitor in the television studio control room that shows the picture going out over the air.

master of ceremony (MC)   See *compere* and *DJ.*

master positive   A laboratory-timed positive film print made from the original negative, used to prepare duplicate negatives. See also *picture master positive* and *sound master positive.*

master shot   See *establishing shot.*

matrix   (1) Electrical signal sources routed and selected through a switching complex. See *routing switcher.* (2) Color filmstrip with gelatin relief images to be combined with two other strips to produce color film in printing.

matte A type of mask, used mainly during the printing process, that blocks off the light in a predetermined pattern in front of the lens.

matte box A square-shaped box mounted in front of a camera lens to hold mattes and filters. Combined with a lens hood it also acts as a sunshade.

MATV Master antenna television.

MAU ITU country code for Mauritius.

MAX Cinemax cable television channel (USA).

maxi brute A high-intensity, long-throw fill light of compact design.

"mayday" Distress signal; CB emergency call.

mb Meter band.

MBC (1) Malawi Broadcasting Corporation. (2) Mauritius Broadcasting Corporation. (3) Middle East Broadcasting Centre (GB), broadcasting via the ARABSAT 1-C satellite (which covers the Middle East and parts of India, to Somalia and the Sudan).

MBNA Min Ben News Agency (Republic of China).

MBS Mutual Broadcasting System (USA).

MC (1) Master control (room). (2) Master of ceremony/ceremonies; compere.

MCO ITU country code for Monaco.

MCR Mobile control room.

MCU Medium close-up.

MCY Magenta, cyan, and yellow in the subtractive color process. See *primary colors/2*.

MDA ITU country code for Moldova.

MDG ITU country code for Madagascar.

MDN ITU country code for Macedonia (provisional code).

MDR (1) ITU country code for Madeira. (2) Mitteldeutscher Rundfunk (Germany).

MDS/MMDS Multipoint distribution service or multiple multipoint distribution service.

M&E track See *music and effects track*.

meat Colloquial term for the principal item (story) in a news program.

mechanical editing See *editing/3*.

MEDIAWAVE International Festival of Visual Arts; film and video festival in animation, art, music, and documentary categories held annually in Győr (Hungary).

medium close-up (MCU) Also called **shoulder shot.** A photographic shot between the close-up and the medium shot.

medium frequency (MF) 300–3,000 kHz.

medium long shot (MLS) A photographic shot between the medium shot and the long shot.

medium shot (MS) A photographic shot that usually covers half of the subject (a person from the waist up).

medium wave European broadcast term for amplitude modulation (AM); 550–1,600 kHz or 187–570 meters. (550–1,720 kHz soon in the USA.)

megahertz MHz; 1,000,000 Hz.

megaphone Also called **bull horn.** A lightweight, compact, funnel-shaped speaker system containing a dynamic microphone, volume control, speaker, and switch for use mostly by the director on location. It can cover an area of 200 to 800 yards (over 50 to 250 m). See also *loud haler.*

Melbourne International Film Festival Annual film festival since 1951 highlighting works of short fiction, animation, documentary, experimental, and student films (Australia).

MENA Middle East News Agency (Egypt).

mercury arc (mercury vapor arc) Arc lamp in which the electrodes are mounted in a small, glass-filled quartz tube. See also *arc/1.*

Merited Artist See *Eminent Artist and Merited Artist.*

mesh effect Mesh beat; an objectionable visible mesh or grill pattern caused by misalignment in an Image Orthicon or Vidicon camera tube. (Now obsolete.)

"Message to Men" International Documentary Film Festival—St. Petersburg Competitive film festival in 16mm and 35mm formats, held in St. Petersburg (Russia).

meter Metric linear measure, 100cm or 1,000mm—equals 3.281 feet or 39.37 inches.

meter band The oldest designation of the internationally (ITU) allocated frequency band, measured in meters.

MEX ITU country code for Mexico.

MF   Medium frequency.

MFD   Minimum focus distance.

MGM   Metro Goldwyn Mayer (USA).

MHz   Megahertz.

mi   Mile.

mic   Abbreviation for *microphone,* especially in Great Britain. See also *mike.*

Mica   Trade name for television standards converter; i.e. to convert PAL video signals to NTSC, or vice versa.

microbar   Also called **barye;** formerly called **abar;** a unit in acoustics, equals one dyne per square centimeter.

microcinematography   Motion picture photography of microscopic objects done usually with a film camera mounted on a microscope.

microcircuit   Miniaturized electronic circuitry employing the integrated circuitry technique.

microgroove   A closely-spaced, V-shaped groove used in long playing (LP) and extended play phonograph records.

micron   A millionth of a meter.

microphone   Also called **mic** or **mike.** A device that transforms sound waves into mechanical energy and electronic impulses. It is the first link in a sound recording system. See also *dynamic microphone* and *ribbon mike.*

microphone boom   See *boom*/1.

microphone stand   An adjustable microphone support placed on the studio floor or table. See *desk stand, floor stand,* or *table stand.*

microwaves   Electromagnetic radiation with very short wavelengths, ranging from 30cm to 1mm, used in line-of-sight wireless transmission. See also *ground waves* and *sky wave.*

mike   Microphone.

mike cable   A specially designed cable connecting the microphone with the recording unit or public address (P.A.) system.

mike clamp   A lightweight, often detachable clamp used to mount a microphone on a stand.

mil   Milli-inch; 1/1000th of an inch, used for measuring tape thickness and width of soundtrack or sound stripe.

mile (mi) British linear measure, 1,760 yards or 5,280 feet; equals 1.609 km (1,609 meters).

milieu See *atmosphere*.

millimeter (mm) 1000th of a meter or 100th of a centimeter; equals 0.0393 inch.

miniature A small-scale model constructed in place of a cumbersome large object to save money and time. It is photographed in such a way that it appears full-scale.

mini brute A compact, portable, high-intensity light used as a filler, or to supplement daylight on location. See also *brute*.

minicam Abbreviation for mini-camera.

mini mac Compact quartz lamp studio light used for broad fill and as a floodlight.

minimum focus distance (MFD) The shortest distance between the object and the lens where the image registers in sharp focus. An important factor for zoom lenses, allowing for a full zoom effect even at close distances.

mini lite A generic term for a variety of small, high-intensity, light-weight studio fill lights; e.g. inkies.

minimum rate The basic fee or payment to players and production personnel allowed by their specific union or guild. See also *flat/4*.

mini-series A television series program running eight or fewer episodes.

Minitel Name of the viewdata terminal in France.

MINPOREN The National Association of Commercial Broadcasters of Japan.

mirror reflex See *reflex camera*.

mirror shot (1) Photographic shot taken of the mirror image of a subject. (2) Shot taken through a suspended mirror over the subject to create a high angle effect. **Note:** mirror shots render a reversed image, i.e. a right-handed violinist, artist, or chef will appear left-handed.

mirror shutter (mirrored shutter) Variable shutter in the film camera that uses mirrors for view-finding purposes.

mi/s (mps) Miles per second.

misframe (1) Incorrect framing of a picture. (2) Wrong splice in a film (miss-frame).

mix   (1) The combination of several soundtracks—narration, dialogue, music, sound effects—into one (mixed) track. See *dubbing/2*. (2) British term for dissolve in a picture.

mix bus   See *bus*.

mixed feed   A picture of one television camera fed into the viewfinder of another for precise superimposition or matting.

mixer   (1) A tape recorder accessory that accepts several microphones. (2) Sound and/or picture control console. (3) Studio engineer handling mix.

mixing panel   See *console/2*.

mix-minus   Minus the commentary; a special audio program feed from a special or sporting event with only the natural sound of the event action and without the commentary by the broadcast or field (stadium) announcer. This audio program feed may be used by the editors with voice-overs, comments, or descriptions during a later broadcast or news program.

mix track (mixed track)   The finished soundtrack on magnetic stock.

MKS system   Meter-kilogram-second system.

MLA   ITU country code for Malaysia.

MLD   ITU country code for Maldives.

MLI   ITU country code for Mali.

MLS   Medium long shot.

MLT   ITU country code for Malta.

mm   Millimeter.

MNA   Myanmar News Agency.

MNG   ITU country code for Mongolia.

mobile unit   Radio or television equipment mounted on a truck or trailer for on-the-scene (remote, OB) broadcast of mostly news, sports, or special events. The sound and picture is relayed back to the studio or to the main transmitter.

Mobilrail   Trade name for studio lighting grid system that works on the principle of roller coupling and is easily movable.

mock-up   A scale model of a set, e.g. the driver's seat of a car, a room, or office corner.

model   A three-dimensional replica of an object or person built to scale for special effects filming.

modeling light Also called **key light.** The principal light source illuminating the subject.

model release A legal agreement by which the studio or production house receives the right to use an individual's voice, picture, and likeness for a show, also for promotion and publicity purposes. In case of a minor, the consent and signature must be given by the parent or legal guardian. See also *property release.*

modulation The change of shape of the carrier wave according to the variations of signals. See *amplitude modulation, frequency modulation,* and *pulse code modulation.*

moire pattern Disturbing effect caused by interference of lines and fine dots in the television system.

Mole (mol) The basic unit of the amount of substance in the SI unit system; a gram molecule.

MOMI Museum of the Moving Image. See *British Film Institute.*

monaural Sound, audio from a single source. See also *binaural* and *quadraphonic.*

monitor (1) High-quality instrument used to check a program for sound and/or picture. (2) A television (picture) receiver in the control room and/or studio. See *jeeped monitor, master monitor,* and *preview monitor.*

monitor head The head on a tape recorder enabling the sound engineer to listen to (monitor) the material during recording.

monitor man See *sound engineer.*

monitor output See *output.*

monochrome (1) Black-and-white picture. (2) B&W signal carried in a color television channel with good resolution.

monopack A reversal type film used in the forties and fifties. See *tripack.*

monophonic recording Sound recording on only one channel.

monopod One-legged camera support for lightweight cameras.

montage (1) A series of abbreviated sequences and musical bridges blended together to indicate the passage of time. (2) Rapidly cut, thematically related film sequences, brief shots, dissolves, and/or superimpositions, used to create a general visual effect. See also *dynamic cutting.*

Moog synthesizer A computerized musical instrument capable of producing a great variety of sounds.

MOS   (1) Mute on sound; a silent film. See *mute*. (2) "Mit-out sound"; American colloquial term attributed to a German film editor.

Mostra Internacional de Cinema de São Paulo   Non-competitive international showing of 16mm and 35mm films and competitive festival for directors up to their first three films. The Bandeira Paulista Trophy (Banner of São Paulo) is awarded. Held in São Paulo (Brazil).

Motion Picture Association of America (MPAA)   Founded in 1922 as a trade association to represent the American theatrical film industry. Today the Association is the advocate and voice for major producers and distributors as well as those involved in television, cable and home video programming. The MPAA provides copyright protection and a copyright advisory, Rating System guidance, advertising administration, technology evaluation and planning, world-wide market research, and title registration. It also acts as a liaison between the government and the film industry and administers the MPAA Archives. See also *Motion Picture Export Association*.

motion picture camera   See *camera/2*.

Motion Picture Export Association of America (MPEAA)   The international arm of the Motion Picture Association of America that provides assistance in international matters that affect the regulations, marketing, sales, leasing, taxation, and distribution of film programs.

motion picture film   A thin, flexible (cellulose derivative) transparent base material coated with a light-sensitive emulsion layer to record photographic images. It is perforated on one or both edges at regular intervals (sprocket holes) to ensure precise movement. Film stocks come in several standard gauges: Super 8, 16mm, Super 16, 35mm, 65mm, 70mm.

motion picture projector   See *projector*.

motion picture studio   See *studio*.

motive   See *theme/1*.

mount   (1) Camera mount. (2) Az/El mount or polar mount that supports a satellite dish.

"Move camera left or right"   See *camera left/right*.

Movietone frame   See *Academy aperture*.

moving coil   See *dynamic microphone*.

moving shot   A scene photographed with the camera in motion.

MOZ   ITU country code for Mozambique.

MP   Motion picture.

MPAA   Motion Picture Association of America.

MPEAA   Motion Picture Export Association of America.

MPEU   Media Program of the European Union. See *EAVE.*

MPMO   Motion Picture Machine Operators (USA).

mps (m/s)   Miles per second.

MR   Magyar Rádió (Hungary).

MRA   ITU country code for the Marianas.

MRC   ITU country code for Morocco.

MRT   (1) Magyar Rádio és Televizió (Hungary). (2) ITU country code for Martinique.

m/s (mps)   Miles per second.

MS   Medium shot.

MSO   Multiple system operator (cable and TV systems).

MSR   ITU country code for Montserrat.

MST   Mountain Standard Time (USA).

MSTV   Association for Maximum Service Television (USA).

MTI   Magyar Távirati Iroda (Hungary).

MTN   ITU country code for Mauritania.

MT&R   Museum of Television and Radio (USA).

MTS   Multi-channel television sound.

MTV   (1) Magyar Televizió (Hungary). (2) Music Television Network (USA).

mug shot   Close-up.

multicam (multiple camera)   Film or film sequence photographed simultaneously with several cameras to avoid retakes and to get a variety of angles and shots for selection in editing.

multi-focus antenna   Also called **spherical antenna.** Antenna with multi-focal points to receive signals from various satellites simultaneously.

multi image   Program presentation; projection of multiple exposures on a single film by one projector. See *multiple image.*

multimedia   (1) Presentation technique, a combination of sound and slide, film and filmstrip and/or live action. (2) New communications medium involving digital sound and interactive techniques in various applications, especially for information and entertainment, i.e. multimedia radio; multimedia television; multimedia records; multimedia print—magazines, books; multimedia advertising; and multimedia telephone.

multiple exposure　Two, three, or more exposures on a single series of motion picture frames.

multiple image　Several images combined and printed on a single piece of film. See also *split screen*.

multiple lens turret　See *lens turret*.

multiple voices　Two or more announcers, often male and female, alternate in presenting a newscast, narration, or voice-over.

Multiplexed Analog Component (MAC)　Satellite television transmission standard utilizing analog picture and digital audio components. The MAC system ensures better use of encryption, thus better protection against program piracy. Variants are:
　In North America, Australia and South Africa: A-, B-, C-, D2-MAC, B-MAC; in Western Europe, Great Britain, and parts of Scandinavia: C-MAC, D-MAC.

multiplexer (MUX)　A set of prisms and movable mirrors for combining images from several sources into one signal channel, as in a film chain.

multiplexing　FM stereo; a form of FM broadcasting in which audio signals in two channels are transmitted on the same carrier, i.e., two or more separate carriers are modulated on the same channel.

Multipoint Distribution Service or **Multiple Multipoint Distribution Service (MDS** or **MMDS)**　Federal Communications Commission-approved television broadcast channels for low-powered short-distance distribution of closed-circuit subscription television programs (USA).

multiscreen　Cinema or filmstrip presentation involving several screens, presented by interlocked (computer controlled) projectors.

multitrack　Multitrack sound or multiple track sound; two or more sound tracks, including a guide track, placed on a release print to provide stereophonic sound.

Munich International Documentary Film Festival　Competitive festival of documentary subjects in 16mm and 35mm films in any length held in Munich (Germany).

Edward R. Murrow Awards   Annual presentation to radio and television news departments for overall excellence in six categories by the Radio-Television News Directors Association (USA).

Museum of the Moving Image (MOMI)   See *British Film Institute.*

musical bridge   See *bridge.*

musical licensing   Authorization of use of, and the collection of fees for, live or recorded music. Licensing is done by the American Society of Composers, Authors, and Publishers (ASCAP) or by Broadcast Music Incorporated (BMI), on behalf of the copyright holder. This may be blanket or per-item licensing. See also *ASCAP* and *BMI.*

musical score   See *score.*

music and effects track   Also called **M&E track.** Sound track containing music and sound effects combined.

music library   One of the most important sections of a broadcast station that contains sheet music, transcriptions, records and tapes, and catalogues filed, cataloged, and handled for the easiest "pull."

music track   Soundtrack containing the musical portion of a program or film.

Mutascope   An early moving picture device, similar to the kinematoscope, used to view rapidly rotating still pictures.

mute   British term for silent film. See *MOS.*

MUX   Multiplexer.

MW   Medium wave.

M/W   Microwave.

MWI   ITU country code for Malawi.

Mylar tape   A very thin, strong, oxide coated, polyester-based magnetic tape, resistant to extreme temperature changes and humidity.

MYT   ITU country code for Mayotte.

N

N   (1) News. (2) Network.

NAB   (1) National Association of Broadcasters (USA). (2) Former News Agency of Burma. See *MNA*.

NABET   National Alliance of Broadcast Employees and Technicians (USA).

NAEB   National Association of Educational Broadcasters (USA).

NAGC   National Association of Government Communicators (USA).

NAMID   National Moving Image Database. See *American Film Institute*.

NANBA   North American National Broadcasters Association (Canada).

nanometer (nm)   One billionth of a meter.

NAPA   North American Photonics Association (USA, Canada).

NAPBC   Native American Public Broadcasting Consortium.

NARMC   National Association of Regional Media Centers (USA).

narration   A voiced description of a film, tape, or sequence read by the narrator and recorded on the narrative (sound) track. See also *commentary* and *VO*.

narrative   The copy, the text of narration.

narrative track   Soundtrack containing the narration for a program.

narrow-angle lens   Telephoto lens.

narrow band   Bandwidth used mostly for police and military communication.

narrow-gauge film   Film format less than 35mm wide, like 16mm or Super 8. See *substandard*.

NASDA   National Space Development Agency of Japan.

NATAS   National Academy of Television Arts and Sciences (USA).

National Academy of Cinematographic Arts and Sciences of Spain
Academia Nacional de las Artes y las Ciencias Cinematografias de
España. See also *Goya*.

National Academy of Television Arts and Sciences (NATAS)
Founded in 1957 for the advancement of the television arts and sciences in news, drama, production, sports, music, film, video and
electronic crafts, writing, directing, design, graphics, performing, education, advertising, management, and support personnel
(USA). The Academy presents the annual Emmy Awards in various
categories, and The Trustees' Award for enduring achievements
and/or contribution to the industry. It publishes the *Television
Quarterly*. See *Emmy Awards*. See also *Academy of Motion Picture
Arts and Sciences*.

National Association of Broadcasters (NAB) An over seventy-year
old trade association representing its members of the radio and
television industries, both nationally and internationally (USA).
The Association provides leadership, legislative, regulatory and
judicial representation, organizes conferences, maintains a library
and information center, and publishes books and relevant information and newsletters, such as *NAB World* for international members, *RadioWeek, TV Today,* and *TechCheck* on technology.

NAB sponsored awards are: the Belva Brissett Award for regulatory achievement; Best of the Best Award in radio promotion;
Broadcasting Hall of Fame in TV and radio; Crystal Radio Award
for community service; Distinguished Service Award; Engineering
Achievement Award, Broadcast; Hugh Malcolm Belville, Jr. Award
in research; Grover Cobb Award for regulatory leadership; Marconi
Radio Awards for programming; National Radio Award for industry
service; Service to Children Award in Television; and AMAX—"AM
at its Maximum" certification.

National Educational Film & Video Festival (NEFVF) Annual showing of educational, special interest and non-theatrical films and
videos in various categories. Awards include plaques, certificates,
and cash prizes for students, and the Gold, Silver and Bronze
Apples (USA).

National Film Awards of India Annual awards covering 20 categories
presented nationally and regionally by the Ministry of Information.
The regional awards are given in 17 languages. The special Phalke
Award honors outstanding contributions in the film industry.

**National Film Board of Canada (NFB); Office National du Film du
Canada (ONF)** One of Canada's national cultural institutions, it
produces and distributes films nationally and worldwide. The films

are produced in English and French by staff and by independent producers. The Film Board maintains production houses in seven locations, and laboratory, post-production, research, distribution and administrative facilities in Montreal and Ottawa.

national news   Newscast covering national events which are usually carried nationwide. See also *international news* and *local news*.

National Radio Awards   See *National Association of Broadcasters*.

National Television Systems Committee (NTSC)   A compatible color television system based on 525-line/50-field (RCA-compatible color system) devised in the USA. It became the standard American color television system in 1953/54. See also *PAL, SECAM* and Appendix E: Television Systems Worldwide.

NATPE   National Association of Television Program Executives (USA).

natural light   Light, the actual daylight provided by the sun. See also *artificial light*.

NAVA   National Audio Visual Association (USA).

NBC   (1) Namibian Broadcasting Corporation. (2) National Broadcasting Company (USA). (3) The former Nigerian Broadcasting Corporation, now FRCN.

NBEA   National Broadcasting Editorial Association (USA).

NBMC   National Black Media Coalition (USA).

NBN   National Black Network (USA).

NBS   Formerly the National Bureau of Standards. See *NIST*.

NCABC   National Catholic Association of Broadcasters and Communicators (USA).

NCG   ITU country code for Nicaragua.

NCL   ITU country code for New Caledonia.

NC17   Theatrical film rating symbol indicating that no one (no children) under 17 will be admitted.

NCTA   National Cable Television Association (USA).

ND   (1) Neutral density; neutral density filter. (2) Non-directional (antenna).

NDR   Norddeutscher Rundfunk (Germany). See *ARD*.

NE   Non-episodic dramatic program.

NEA   National Endowment for the Arts (USA).

near-ultraviolet light   See *black light.*

neck microphone   See *lavalier.*

needle   Turntable arm pick-up needle, stylus.

needle drop   One-time use of copyrighted, licensed music, musical piece, or composition.

NEFVF   National Educational Film & Video Festival.

negative   (1) Negative (reversed) image. (2) Negative raw stock, or camera film, as opposed to positive film. It must be printed on another stock for viewing. (3) The original, exposed negative film. (4) Processed film with a negative image. See also *positive* and *reversal.*

negative assembly (negative cutting)   Also called **confirming**; the cutting and editing of the original negative film to match the approved workprint.

negative perforation   Also called **BH standard.** Standard size and form of sprocket hole of 35mm negative film and intermediate stock. Negative perforation is designed to render a steady image during exposure and intermittent movement of the film in the camera. See also *KS Standard* or *positive perforation.*

negative-positive (negative-positive process)   A positive image is obtained by developing the latent image by way of printing from a negative. This process, the basis of modern photography, was introduced by Henry Fox Talbot in England in 1839.

negative projection (phase reverse)   Polarity of the video signal reversed by circuitry. The tones are inverted, turning the negative into positive or the positive into negative. This type of projection might damage the original film and may be used only once.

Nellie   Name of the annual award sculpture presented by ACTRA—the Alliance of Canadian Television and Radio Artists, in approximately 16 categories, since 1972.

nemo   Field pick-up, remote (an archaic term).

NESN   New England Sports Network pay television (USA).

NET   (1) Network; a series of broadcast stations connected by lines or satellite(s) that carry a program simultaneously. (2) National Educational Television (USA); now defunct. See *PBS.* (3) Diffusion material (net) hung in front of a light.

neutral density (ND)   Also called **ND filter.** Neutral density filter; a camera filter that reduces the intensity of light without altering its characteristics.

neutral shot See *cutaway*.

neutral test card See *gray scale*.

news The reporting of information about current, actual events.

news bureau A designated office of a large network or station away from the home base from where correspondents compile and regularly send news items.

news camera (1) A compact, portable, self-contained television camera used for ENG—electronic news gathering and/or EFP—electronic field production. It may record the events on videotape or relay them directly to the station (control) via microwave or satellite. (2) A compact, silent (old system), or single-sound system, portable film camera equipped for flexible and fast operation without cumbersome lighting and set-up.

newscaster The announcer or reporter of broadcast news.

newsfilm A short videotape or film covering current events, prepared and edited for news broadcast. It may be a silent, or a sound clip, or a narrated segment with an on- or off-camera reporter, or it may be an interview sequence.

news flash Short, periodic announcement of an ongoing (news) event.

news program A popular broadcast program compiled from several news items and from different sources, where speed, fast assembly of material and equipment and up-to-the minute information are essential.

newsreel (newsfilm) Short film of current events shown mainly in cinemas.

news service News wire service.

New York Film Festival Invitational, non-competitive, yearly film festival sponsored by the Film Society of Lincoln Center for independent short and feature films of foreign and American productions. Certificate of Participation is awarded to selected films (USA).

N.F. (1) Normal frequency. (2) Nominal frequency.

NFB National Film Board of Canada.

NFBD National Film Board of Denmark.

N.G. No good; term used for film rejects. See *bad footage* and *bad take*.

NGR ITU country code for the Republic of Niger.

NH Normal hours.

NHK Nippon Hoso Kyokai.

NiCad Nickel cadmium, heavy-duty, portable, rechargeable battery. See also *battery.*

NICK Nickelodeon cable television channel (USA).

nickelodeon A movie house during the early twentieth century that charged 5 cents (a nickel) for admission. Also a **player piano** or a **juke box.**

NIG ITU country code for Nigeria.

night effect Also called **day-for-night.** A technique used in cinematography to create the appearance of night by the use of filters while the exterior scene is actually filmed in daylight.

NII National Information Infrastructure.

9.52cm/s Recording tape speed: $3\frac{3}{4}$ ips.

9.6 ips Magnetic (professional) videotape speed: 24.3cm/s.

19.05cm/s Recording tape speed: $7\frac{1}{2}$ ips.

19.05mm Magnetic tape width standard, approximately ¾ inch.

Nipkow disc Nipkow's mechanical scanning wheel; late nineteenth-century television scanning device with perforated holes. Named after Paul Nipkow (1860–1940), German engineer.

Nippon Academy Awards Yearly Academy Awards of Japan since 1978, presented by the Nippon Academy-Sho Association in fourteen categories for Japanese feature films and one foreign film category. The annual awards presentation takes place in March. See also *Geijutsu Sakuhin Sho.*

Nippon Academy-Sho Association Japanese non-governmental organization with a membership of over 5,100 film and film-related professionals, sponsor of the Nippon Academy Awards.

Nippon Hoso Kyokai (NHK) Japan Broadcasting Corporation.

NISO National Information Standards Organization (USA).

NIST National Institute of Standards and Technology; formerly NBS-National Bureau of Standards (USA).

nitrate Cellulose nitrate base; a flammable material used in film manufacturing. Replaced over 30 years ago by safer materials. See *nonflam film* or *safety film.*

NIU ITU country code for Niue Island.

NMB ITU country code for Namibia.

NOB   Nederlands Omroep Bedrijf; Netherlands Broadcasting Service.

nod shot (nodding shot)   A silent reaction shot of a listener, host, or interviewer. See also *reaction shot.*

noise   (1) Any unwanted sound in acoustics. (2) Undesired electronic signal. See also *snow.*

noiseless camera   Silent, self-blimped film camera.

nomography of filter   The calculation (chart) of the filter's effect on color temperature.

non-broadcast use   Communication like CB, ham radios, corporate radio dispatch, and cellular telephones, not broadcast for the general public. See also *carrier current.*

non-directional microphone   See *omni-directional mike.*

non-flam film   Also called **safety film.** Non-flammable film; a film base with a slow burning rate.

non-program material   Commercial announcements, advertising material.

noodling   The playing of a few bars of the background music while the titles are rolling.

nook light   Colloquial term for a small broad light used in confined areas.

NOR   ITU country code for Norway.

NOREA   Nordic Radio Evangelistic Association (Norway).

normal lens   A general term for non-telephoto or non-wide-angle lens. In old-style television cameras (1948–64), equipped with turrets, a 75mm or 90mm lens was considered "normal." In 35mm film cameras it is usually a 50mm (2 in.) lens; in 16mm cameras it is a 25mm (1 in.) lens.

NOS   Nederlandse Omroep Stichting; Netherlands Broadcasting Corporation.

notch   Small cut on the edge of the negative film to actuate, cue, light changes in printing.

note   A musical tone (sound) of definite pitch and length.

NP   Notícias de Portugal.

NPA   Namibian Press Agency.

NPC   National Press Club, founded in 1908 (USA).

NPL ITU country code for Nepal,

NPR (1) National Public Radio (USA). (2) Noiseless portable reflex (camera).

NRB National Religious Broadcasters (USA).

NRBA National Radio Broadcasters Association (USA).

NRK Norsk Rikskringskasting (Norway).

NRU ITU country code for Nauru.

NTA Nigerian Television Authority.

NTB Norsk Telegrambyra (Norway).

NTC National Telemedia Council (USA).

NTSC National Television Systems Committee (USA). See also *PAL* and *SECAM*.

NTV The Nippon Television Network Corporation (Japan).

number card A cue card with numbers on it indicating time or other numbered cues.

numbering machine A machine used in film laboratories for printing edge (key) numbers along the edges of film at regular intervals.

nx News.

NZBC New Zealand Broadcasting Corporation.

NZL ITU country code for New Zealand.

NZPA New Zealand Press Association.

NZWG The New Zealand Writers Guild.

OB   Outside broadcast. See *remote*.

Obie light   See *camera light*.

objective   The component in an optical system, i.e. lens, that forms an image.

OB van   See *remote van*.

OC   Open-captioned; television viewing symbol, visible without a decoder.

OCE   ITU country code for (French) Polynesia.

OCV   On-command video.

off-camera   (1) The performer or subject out of the camera's viewing range. (2) Narrator or commentator heard, but not seen. See *VO*.

Office National du Film du Canada (ONF)   See *National Film Board of Canada-NFB*.

off-line editing   Off-line videotape edit; the first part of videotape editing, similar to the workprint in motion picture film, where assembly and shot decisions are made. See also *on-line editing*.

off-mike   Performer speaking away from the microphone either to create a desired effect, or as a result of a mistake.

offset   See *interlace*.

offset title   See *drop shadow*.

offstage   Off-camera.

off-sync (out of sync)   Synchronization that is out of line.

ohm   The unit of electrical resistance and impedance. Electromagnetic force V (volts) = A (amperes) × R (resistance in ohms). Named after Georg Simon Ohm (1787-1854), a German physicist.

OIRT   The former Organisation Internationale des Radio et Télévision; International Radio and Television Organization-IRTO. Now *EBU*.

OITS Organisation Internationale de Télécommunicacions par Satellites; Organizacion Internacional de Telecommunicacion por Satellite. See *Intelsat*.

OMA ITU country code for Oman.

OMC Operations and Maintenance Center.

omega loop/wrap Magnetic tape wound around the drum in the shape of the Greek letter Ω. See also *alpha loop* and *half loop*.

omni-directional mike Also called **multi-directional mike.** A non-directional microphone capable of picking up signals from all directions. See also *bi-directional mike* and *uni-directional mike*.

on-board battery Snap-on battery; a small battery or battery pack attached directly to the camera. See *battery*.

100 ft Standard film reel size containing 100 feet, or appr. 30 meters of film.

122 m Standard film reel size containing 122 meters or 400 feet of film.

1 inch (1) Magnetic (video) tape width standard—equals 25.4 cm. (2) One inch lens—25mm.

1.85:1 Also called **one eight five.** Aspect ratio in normal screen projection—4 × 3. 1/30 and 1/25. See also *2.35:1*.

one-light print A quickly produced workprint and subsequent copies of it, made with a single printer light setting.

one-light videotape A direct tape-to-tape transfer copy with no corrections.

1⅞ ips Magnetic tape speed standard that equals 4.76 cm/s.

one-shot A single performance or program not scheduled for repeat.

1,125-line/60-field High-definition television (HDTV) system.

1,200 ft Standard film reel size containing 1,200 feet or appr. 366 meters of film.

one-to-many broadcasting See *point-to-multipoint*.

One-to-one broadcasting Also called **pointcasting.** Point-to-point broadcasting.

ONF Office National du Film du Canada. See *National Film Board of Canada-NFB*.

on-line editing (on-line videotape editing) The final stage of editing, resulting in a broadcast quality videotape, like a motion picture film master. See also *off-line editing*.

ONPTZ   Office des Postes et Télécommunications du Zaire.

on-the-air   Broadcast program in progress.

on-the-cuff   A performance with no remuneration or payment.

"on-the-nose"   Colloquial expression for a program carried out exactly on time.

O&O   Owned and operated; a broadcast station owned and operated by a network.

OOV   Out-of-vision. See *off-camera/2*.

OP   Orient Press (Republic of Korea).

opacity   See *density*.

opaque leader   A transparent, off-white film leader used in the editing process, especially for markings.

open-air theater   See *drive-in theater*.

open end   A broadcast program that allows time for local announcements and station ID.

open-ended film   Film program of a mostly educational nature that allows for spontaneous response. See *single-concept film, educational broadcasting/film*.

open-end net   Net placed in front of a light to reduce its intensity and/or render light balance. See also *black net* and *lavender net*.

open mike   Also called **hot mike** or **live mike**. A microphone turned on.

open reel   Tape or film reel of standard size not encased in a cassette or container.

open set   A set not closed or boxed in by walls.

opening shot   See *establishing shot*.

operator   Technician operating the camera, projector, or printer or other equipment. See also *camera operator* and *projectionist*.

OPI   Office of Public Information of UNESCO. See *United Nations Educational, Scientific and Cultural Organization*.

optical effects   Also called **visual effects** or **opticals.** (1) Alterations effected by electronic means (computer) in television. (2) Alterations, like dissolves, fades, hold-frames, and wipes in film, usually made in an optical printer. See also *sound effects* and *special effects*.

optical enlargement   Also called **print up.** A small gauge negative film enlarged in the laboratory to a larger format, i.e. 16mm to 35mm.

optical/magnetic projector   Sound film projector capable of projecting films with either optical or magnetic sound tracks.

optical printer   Printing machine in a film laboratory that prints through a lens system to permit reduction, enlargements, or optical effects. See also *contact printing* and *step printer.*

optical recorder   A device used to transfer the audio information from an electrical signal to optical image.

optical reduction print   See *reduction print.*

opticals   See *optical effects.*

optical sound   Sound recording method in 35mm and/or 16mm film production using the variable area or variable density method. Seldom used in original recordings.

optical sound head   Electrical device for the reproduction of optically recorded sound-on-film to electrical signals.

optical track   Also called **photographic soundtrack.** An optical soundtrack recorded by light modulations of variable-area or variable-density method. See also *soundtrack/1.*

OPTT   Office des Postes et Télécommunications du Togo.

orange filter   Also called **#85** or **#85B filter**. Orange conversion filter used to transform daylight into tungsten color temperature, i.e. an #85 filter for 5400K to 3200K conversion or an #85B filter for 5400K to 3400K conversion. See also *color conversion* and *blue/2.*

ORB   Ostdeutscher Rundfunk Brandenburg (Germany). See *ARD.*

orchestra box   Orchestra pit.

ORF   Österreichischer Rundfunk (Austria).

Organisation Internationale de Radio et Télévision (OIRT)   International Radio and Television Organization (IRTO). Now *EBU.* See *European Broadcasting Union.*

Organizacion de la Television Ibero Americana (OTI)   Ibero-American Television Organization (Mexico).

orientation shot   See *establishing shot.*

original   (1) The first recording of a sound or videotape. (2) The film exposed in the camera.

Orion   International satellite of British Aerospace.

ORT/ORTF   Office de Radiodiffusion Télévision Française until 1974. See *TDF* and *TF*.

ORTB   Office de Radiodiffusion et Télévision de Benin.

Orthicon   See *Image Orthicon*.

orthochromatic film   Ortho film; B&W film emulsion with sensitivity to only blue and green light.

ORTN   Office de Radiodiffusion-Télévision du Niger.

ORTO   Olympics Radio Television Organization.

Orwocolor   A negative/positive integral tri-pack color film process, similar to *Agfacolor,* that was manufactured in the former Democratic Republic of Germany. Now obsolete.

Oscar   The Academy Award; popular name of the golden statuette awarded yearly and presented by the Academy of Motion Picture Arts and Sciences (USA). The 13½-inch, 8½-pound statuette was designed in 1927 by MGM art director Cedric Gibbons, sculpted by George Stanley, and cast in an alloy of tin and antimony, then copper-, nickel-, silver-, and gold-plated by the A. J. Bayer Metal Co. It represents a naked knight with a sword on a five-spoked film reel depicting the original five branches of the Academy: actors, directors, producers, technicians, and writers.

In 1931 Margaret Herrick of the Academy commented that it looked like her uncle Oscar Pierce of Texas. The nickname stuck. See *Academy Awards* and *Academy of Motion Picture Arts and Sciences.* See also *Emmy*.

oscillator   An electronic circuit producing continuous oscillations at fixed or adjustable frequencies.

oscilloscope   A measuring device showing electronic patterns by way of a cathode ray tube.

OTE (ET)   Elliniki Tileorassi (Greece).

OTI   Organizacion de la Television Ibero Americana; Ibero-American Television Organization (Mexico).

Ottawa International Animation Festival   Biannual international festival for animated films in various categories organized by the Canadian Film Institute and held in Ottawa (Canada).

out   (1) Out-cue; the closing word(s) of a story or program. See also *IN/2*. (2) Out point; the end of an edit.

out cut   See *out-takes*, and *cut/5*.

outdoor　See *exterior*.

outline　Main aspects or essential features of a story intended for a teleplay or a film script, written before the synopsis. See *script, synopsis, teleplay*.

out of focus　Unclear, blurred image caused by incorrect focusing in the camera or projector.

out of frame　(1) See *Off-camera*. (2) Parts of two frames appear on the screen, a result of faulty projection alignment.

out of sync　See *off-sync*.

output　Connection or terminal for signal voltages coming from electrical components such as amplifiers and preamplifiers. Tape recorders have line, monitor and speaker outputs. See also *input*.

outs　See *out-takes*.

outside broadcast (OB)　British term for remote broadcast.

out-takes　Also called **outs.** Rejected (N.G.) or unused takes from the final, edited version of a film. See also *bad footage* and *bad takes*.

out time　The exact time of ending the show.

over cranking　See *high-speed camera*.

overexposure　Exposure of the film in the camera with a great light intensity, resulting in an unsatisfactory tonal range and loss of detail. In the case of color film it can give distorted color rendering, a dense negative, and washed-out positive.

overlap　Editing the dialogue in such a way that the sound is extended over the next shot, e.g. the picture shifts from the interviewer to the subject while the former's voice is still being heard.

overlay　Superimposing technique of the television picture. See also *inlay*.

overrun　To exceed the allocated (out) time.

over-the-shoulder shot　Reverse-angle shot of a subject, performer, onlooker, photographed from an over-the-shoulder point of view. See also *reaction shot*.

OWI　Office of War Information. See *United States Office of War Information*.

owned and operated (O&O)　A broadcast station owned and operated by a network. See also *affiliate* and *independent broadcast station*.

oxide　See *iron oxide*.

P

P   Program.

PA   (1) Performing arts. (2) Power amplifier. (3) Production assistant. (4) The Press Association (GB).

P.A.   Public address (loudspeaker) system.

pace   The speed, the tempo of the overall performance.

package   A broadcast program or show prepared (usually by an agency and/or production company) and sold in its entirety to a station or advertiser.

painted matte   A painted scene or background that is photographed and combined with live action.

PAK   ITU country code for Pakistan.

PAL   Phase Alternating Line. See also *NTSC* and *SECAM*.

Palapa   Name of Indonesian satellite.

Palm d'Or   Golden Palm; award of the yearly Cannes International Film Festival. See *Festival International du Film-Cannes*.

PAN   See *panoramic shot*.

Panaflasher   Device used for simultaneous flashing on *Panavision* film cameras. It may be used before or after exposure.

Pan African Film and Television Festival of Ouagadougou   See *FES-PACO*.

PanAmSat; Pan American Satellite-(Alpha Lyracom)   An international commercial satellite consortium serving the United States, Latin America, and Europe, based in Greenwich, Connecticut (USA).

PANARTES   Pan American Federation of Arts, Mass Media and Entertainment Unions.

Panavision   Trade name for wide-screen processes: (1) *Panavision*; the film gauge is 35mm, but an anamorphic lens is used on the camera and on the projector. (2) *Super-Panavision*; the original is 65mm film, but the prints are made on 35mm by anamorphic reduction. (3) *Ultra-Panavision*; a ratio of 1.25:1 is achieved by photographing with an anamorphic lens on 65mm negative film. See also *Cinemascope*.

panchromatic (pan film)   B&W camera film with sensitivity to all colors of the visible spectrum. See also *color film*.

panchromatic viewing filter   A filter used by the cinematographer to determine how a scene will be rendered on panchromatic (B&W) motion picture film and to check relative brightness and lighting contrast. See also *color contrast viewing filter*.

panel   (1) See *board*. (2) Guests on an interview, talk, or discussion type program.

pan film   See *panchromatic*.

pan handle   Also called **panning bar.** A handle with (usually) a screw mount fixed to the pan head to facilitate movement.

pan head   Tripod head. See *fluid head, friction head,* and *gyro head*.

panning   See *panoramic shot*.

panning bar   See *pan handle*.

panoramic shot (PAN)   The horizontal (panning) movement of the camera, either on a tripod or hand-held. See also *tilt*.

pantograph   An expandable device used for hanging studio lights. Called a **lazy boy** in Great Britain.

PAP   Polska Agencja Prasowa (Poland).

parabole   A microphone with a parabolic mounting used for long-distance sound pick-up. See *long-range microphone*.

parallax   The difference as seen through a side- (or top-) mounted, non-reflex viewfinder and through the camera lens. Cameras with separate viewfinders are equipped with parallax correction.

par-mac   A multiple purpose, weather resistant light for large area and/or location illumination.

PAS   Pan-American Satellite Corporation (USA).

pass band   A definite band that freely transmits signal frequencies in a selective network.

passive satellite   Communication satellite that only reflects the transmission from its surface. See also *active satellite*.

P.A. system Public address system; microphone, amplifier, loud-speaker.

PAT Press Association of Thailand.

patch cord A short cable with plugs at either end used to interconnect equipment, usually at closely grouped connection jacks that are input and output.

patching The emergency interconnection of equipment by a short patch cord.

pay (pay TV) Pay television; cable television service designation for Home Box Office, Cinemax, Showtime, The Movie Channel, Disney Channel, and the Sports Channel.

PBC Pakistan Broadcasting Corporation.

PBS Public Broadcasting Service (USA). See also *NET/2*.

PBY Playboy pay television network (USA).

PC Printed circuit.

PCM Pulse code modulation.

PCSC Pontifical Council for Social Communications (Vatican City State).

PD (1) Public domain; in the public domain. (2) Program director.

Peabody Award George Foster Peabody Award. Annual presentation in six categories for achievements in radio, television, and cable television by the Henry W. Grady School of Journalism, University of Georgia. (USA)

PE/AFL-CIO Professional Employees Department of the American Federation of Labor—Congress of Industrial Organizations.

peak The relative volume of transmitted sound; the highest point reached by the volume indicator.

peak black See *black level.*

peak program meter VU meter to measure peak values of sound signals, used in Great Britain.

peak time Also called **Class A time** or **prime time**. Time period during which the largest audience is listening and/or viewing; the four evening hours: 7 p.m.–11 p.m. (6 p.m.–10 p.m.).

peak white See *white level.*

pedestal (1) A hydraulically operated telescopic dolly with a column-type mount for a television camera, facilitating easy up-and-down movement. (2) The indication of the black level of a television picture.

PEG   Public access channel and bulletin board.

pegs, pegs track   See *pins*.

P.E.N.   P.E.N. Club; International Club of Poets, Playwrights, Essayists and Novelists.

perambulator   A three- or four-wheel dolly on which a microphone boom or camera can be placed.

perforation   Also called **perf** or **sprocket holes.** Film perforation; holes accurately spaced along one or both edges of the motion picture film, ensuring precise location and movement in the film mechanism. See *double perforation* and *single perforation*. See also *BH standard* and *KS standard*.

performer   Actor, talent.

periscope finder   A camera viewfinder that can be rotated to give an all-around oblique view.

permission   See *copyright, license, model release, property release*.

persistence of vision   A characteristic of the human eye by which the retina retains the image of an object an instant after it has been removed. Successive images, if they follow one another in a rapid pace, produce a continuous impression. This phenomenon permits the illusion of movement in cinematography.

personal microphone   See *lavalier*.

perspective   See *sound perspective*.

Perutz   Perutz film; tri-pack color film based on the *Agfa* process. It is now *Werk Perutz Magnetband Audio Video of Agfa-Gevaert* (Germany).

PEVE   Prensa Venezolana (Venezuela).

PG   Film rating indicating parental guidance is suggested.

PGA   Producers Guild of America.

PG13   Film rating indicating that parents are strongly cautioned; no one under 13 years of age will be admitted without a parent or guardian.

PH   Phase.

Phalke Award   Indian governmental award to honor outstanding contributions in the film industry. See *National Film Awards of India*.

Phase Alternating Line (PAL)   A color television standard based on the NTSC system, with a 625-line/50-field, developed in Germany by Dr. W. Burch and used in the European-Continental system. It is the most widely used color television system in the world. See also *NTSC, SECAM,* and Appendix E: Television Systems Worldwide.

phase-reverse   See *negative projection.*

phasing   Proper polarity orientation of two speakers in a stereo playback with a common ground connection to both speakers.

Phenakistoscope   A viewing device consisting of two discs mounted on a shaft. One disc carried the drawings which were viewed through slits around the outer edge of the other disc. It was invented in 1833 by Joseph Antoine Plateau (1801–1883) in Ghent, Belgium. Simon Ritter von Stampfer of Austria also created a similar device.

PHL   ITU country code for the Philippines.

phot (ph)   A unit of illumination, equals 1 lumen per one square centimeter.

photocell   Device that converts light intensity into electrical signals.

photoflood   Incandescent tungsten light bulb with an increased light output, and a relatively short life.

photographic density   See *density.*

photographic lighting principle   Basic lighting in both television and motion picture production using a key light, fill light, and back light. See *lighting.*

photographic revolver   Also called **photographic gun.** The first self-contained camera, in a large revolver shape, capable of taking pictures in rapid sequences, developed by Etienne Jules Marey (1830–1904), French physiologist. The barrel acted as a lens and the film strip was contained in the rotating chamber.

photographic soundtrack   See *optical track.*

photometer   See *exposure meter.*

photometric filter   See *daylight conversion/2.*

PIBA   Pacific Islands Broadcasting Association (Vanuatu).

pick-up   (1) The location of a remote (OB) program. (2) The reproduced sound or visual transmission as a result of the placement of microphones and cameras. (3) Neutral, cut-away film shots, i.e. pick-up shots. (4) The arm of a phonograph or transcription. (5) To pick up; to increase the pace, tempo, or performance.

pick-up tube   See *television pick-up tube*.

picture   (1) The visual image of television or film. (2) The motion picture film program itself.

picture area   Full frame.

picture composition   An esthetic arrangement of a scene or scenery composed and photographed in a pleasing manner.

picture frequency   See *frame*/1.

picture head   See *head*/3.

picture-in-picture (PIP)   More than one television picture from multiple sources shown on the receiver screen.

picture master positive   A good quality positive print derived from the original negative, used for making dupe (duplicate) negative.

picture negative   See *negative*/2.

picture noise   See *noise*/2. See also *snow*.

picture registration   See *registration*.

picture tube   Cathode ray tube in a television receiver.

pilot   (1) The first (demonstration) program or film of a proposed series. (2) A continuously transmitted signal providing a constant system check. (3) Color film laboratory test strip. See *test*/4.

pilot beam   Electromagnetic beam in certain color receiver tubes guiding the writing beam for correct color reproduction. See also *writing beam*.

pilot pin   See *pins*.

pilot print   Short strips of color film supplied with a B&W workprint to show quality of color.

pilot tone   A control track recorded in addition to the actual recording on a magnetic sound recorder to ensure synchronization with the camera or projector. Used in double system remote recording.

pin   See *pins* and *beep*.

pinch roller   See *pressure roller*.

pins   Also called **pegs** or **registration pins.** Fixed positioning pegs in a film camera, printer, projector, or animation stand that engage the perforation holes to provide exact placement of the film or animation cel, with reference to picture aperture.

PIP   Picture-in-picture.

pipe grid   See *grid*.

piping   See *feed*.

pistol grip   A device in the shape of a pistol handle attached to equipment (camera, shoulder pod, long range mike, bullhorn, etc.) to provide a firmer grip in hand held operations. The pistol grip can be equipped with trigger controls and start/stop (on/off) switches.

pit   Orchestra box in theaters.

pitch   (1) The degree of acuteness or accuracy, the quality of sound, measured in Hz units. (2) The distance from the bottom edge of one perforation to the bottom edge of the next in 35mm film. (2/a) The distance between the center of two successive frames in 16mm film.

   Normal-speed film cameras require films with short pitch. Short pitch negative film is also used in contact printing. High-speed film cameras require films with long pitch.

pitch advertising   Excessively long commercials.

pix   Colloquial for *pictures*.

pixels   Picture elements. In a projected 35mm film frame the number of pixels is appr. 1 million. To compare, in a home television receiver frame, 150,000 pixels are displayed.

pixillation   A distortion of the picture clarity for desired effect. Single-frame animation technique developed by Norman McLaren of the National Film Board of Canada.

PL   Principal language.

platter   A phonograph record, transcription, or disc.

playback   (1) The immediate replay and monitoring of a sound or video recording. (2) The replay of a previously recorded song or musical number in a studio for synchronization purposes with acting and dance.

playback head   The magnetic head that plays back the signals from a tape.

plot   The story line, the plan, the outlined pattern of the action, narrative, or drama of a radio play, teleplay or screenplay.

plug   (1) A mechanical interconnector used for easy connection of components. (2) Trade slang for a short promotion, a commercial announcement; also an ID of donors of game shows.

Plumbicon   Trade name for a photoconductive "lead oxide" television camera tube used in modern television cameras. See also *Image Orthicon, Vidicon* and *Saticon*.

PM (1) Prime meridian. (2) Post meridian (USA). (3) Production manager. See *Unit production manager.*

PMPEA Professional Motion Picture Equipment Association (USA).

PNG ITU country code for Papua New Guinea.

PNR ITU country code for Panama.

PNS Philippine News Service.

PNZ ITU country code for Panama Canal Zone.

P.O. Present operation.

pod Grouped, a group of advertisements or announcements placed in a broadcast program.

pointcasting See *point-to-point.*

point-of-view camera Also called **lipstick camera.** A micro video camera, appr. 3 to 4 inches long (the size of a lipstick tube) attached to a race driver's, skier's, or jockey's helmet or concealed in other places to give the point-of-view of the actor. Due to its small size and ease of concealment, it is used extensively in news investigative reporting. See also *video sunglasses.*

point-to-multipoint Also called **one-to-many broadcasting.** Transmission from one source to multiple receivers.

point-to-point Also called **pointcasting.** Point-to-point transmission; broadcasting from a source to an intended (designated) receiver.

POL ITU country code for Poland.

polarization Antenna polarization; modification of vibrations of wave motion so that the ray exhibits different properties on different sides—opposite sides being alike, and those at right angles showing maximum difference. Polarization may be vertical or horizontal.

polarizing filter A specially treated filter made of synthetic material used for reducing water reflections, glare, or rays of polarized light from the sky.

polar mount See *mount/2.*

polar orbit The path (orbit) of a satellite perpendicular to the plane of the equator.

poly-directional mike Also called **omni-directional mike.** Multi-directional mike; microphone with adjustable pick-up areas.

polyester base film See *safety film.*

POR   (1) ITU country code for Portugal. (2) Pacific Ocean Region; satellite ocean region specification. See *ITU region.*

portable camera   Also called **field camera.** A lightweight, self-contained video or film camera, used in ENG, ESG and remote (OB) filming. See also *studio camera.*

portable dolly   A three-wheeled dolly that folds into a compact unit for easy transport.

ported speaker   See *tuned speaker.*

positioning   The placement of microphones and cameras for the best possible sound and/or visual pick-up.

positive   (1) The positive image; the true image of the original subject or scene. (2) A positive film, a reversal type film stock. Also called **reversal.** (3) A positive print made from a negative film; a master positive or a release print. See also *negative.*

positive perforation   KS Standard; standard form and size of the sprocket hole on 35mm positive film stock, also on 65mm negative and 70mm positive film. Positive perforation is designed with a shape to reduce cracking and film damage during repeated projection. See also *BH standard* or *negative perforation.*

post-flashing   Neutral exposure used to reduce contrast and increase shadow detail in film. See also *pre-flashing* and *Panaflasher.*

post-production   Work and activity following the filming process. This includes picture and sound editing, music, effects, printing, and other arrangements to complete the program or film. See also *pre-production.*

post-recording   Sound recording done after the completion of the visual part.

post-synchronization (post-sync)   Recording and adding sound to a picture after the filming has been completed in a manner that the action coincides with the sound (effect, music) and dialogue. See also *dubbing/3.*

POT   Potentiometer. See *fader.*

POV   Point-of-view. See *point-of-view camera.*

power   (1) Electrical power, "juice." (2) Power of lens. See *f-number.* (3) Power of station.

power amplifier (PA)   Amplifier designed to operate a power system.

power cable/cord   Cable used to connect equipment with the power source, i.e. with a battery or AC current.

power of station The power, expressed in watts, allocated for program transmission to a broadcast station. Called simply: Power.

power pack See *battery*.

power rail Rail fixture (with restricted flexibility) for studio lighting that carries uninterrupted power while lamps are being moved.

power zoom Motor driven zoom lens.

PPI Pakistan Press International.

PPM Peak program meter.

ppm (p/m) Pictures per minute.

pps (p/s) Pictures per second.

PPV Pay-per-view cable or DBS television.

PR Public relations.

Praxinoscope Mirrored drum, an early animation device, developed in 1877 by Emile Reynaud (1844–1918), French inventor.

PRCS Personal radio communications service.

pre-amplifier The first stage of amplification that boosts the extremely weak signal voltages (from microphones, magnetic playback heads, or turntable pick-ups) to a level usable by power amplifiers, and equalizes levels to acceptable standards.

pre-empt The replacement of the regularly scheduled program with an unscheduled special event or priority program.

pre-emptibility Advertising term for a less expensive spot that may be pre-empted for a higher paying client.

pre-flashing The uniform exposure of the camera film to a small amount of light before its use to reduce its contrast and increase shadow detail. In general, however, post-flashing is preferred. See *post-flashing* and *Panaflasher*.

PRELA Prensa Latina (Cuba).

premier First; the first public presentation of a theatrical (feature) film, play, or other performance, preceded usually with a great deal of promotion and publicity.

premier plan French for *foreground*. Also *close-up (CU)* in parts of the Continent. See also *gros plan*.

Premio Nacional National Awards; yearly presentation certificate and monetary awards for cinematography, the arts, and the media by the Ministry of Culture of Costa Rica.

pre-production   Preparation and planning activity for a program or film before the start of production.

prequel   New television episode or movie preceding the events of the original episode; the opposite of *sequel*.

pre-recorded   A program recorded on audio- or videotape (television) prior to its broadcast.

pre-recording   The recording of the sound portion of a video program or film prior to taping the visual portion.

pre-score   Music composed and recorded prior to filming or taping a program.

presence   (1) Microphone pick-up that is finely projected and has the air of intimacy. (2) The bearing, the manner of an actor/actress and the atmosphere he/she is able to create on the set or stage.

presentation   The technique of presenting a program in continuity, including stand-by programs, breakdowns, emergencies, apologies, etc.

press or **publicity release**   News or promotional information released to the media. See also *release*.

pressure mike   See *dynamic microphone*.

pressure pad   Felt pad mounted on an arm in recording/playback equipment to hold the tape in close contact with the (recording/playback) heads.

pressure plate   Device in the film camera, projector, or optical printer that keeps the film in place, i.e. in the focal plane of the lens.

pressure roller   Also called **pinch roller.** A roller, usually made of rubber, in recording/playback equipment, that engages the capstan and pulls the tape across the heads at constant speed without slippage. See also *capstan*.

pre-striped   Pre-striped film; magnetic stripe applied to the edge of the film in manufacture.

preview   (1) Previewing a performance, program, or sections of it before selection for transmission. (2) The showing of a motion picture film or videotape program to an invited, selected audience— usually the press, critics, and distributors—prior to its release to the general public.

preview monitor   A control room picture monitor in a television studio used for program selection and technical evaluation.

PRG   ITU country code for Paraguay.

PRI Public Radio International; formerly *APR*.

primary affiliate Broadcast station that is a sole or main affiliate of a network. See also *secondary affiliate*.

primary area The area of a broadcast station where its transmitted signal comes through regularly.

primary colors (1) Red, green and blue in the three-color additive process. (2) Magenta (reddish-blue), cyan (blue-green), and yellow pigments in the subtractive color process.

primary movement The movement in front of the camera. See also *secondary movement*.

prime focus antenna See *center focus antenna*.

prime lens Lens with a fixed focal length.

prime time (PT) See *peak time*.

Prime Time Access Rule See *access time*.

print See *positive/3*.

printed circuit (PC) Compact, "printed" wiring of thin copper foil over laminated material used in modern sound and video/television equipment for interconnection of various circuit components.

printer (1) Machine in the motion picture laboratory that makes prints from the original exposed camera film. The types of printers are *contact, optical,* and *step*. (2) The technician operating the printing machine.

print film Film stock used for printing the workprint and copies of the original (edited) film. See also *camera film* and *laboratory film*.

printing (1) The process of printing. (2) Print-through. (3) The copying of motion picture images onto film by photographic process.

"Print it" See *"It's a print."*

print-through Also called **printing.** (1) Unwanted sound and/or background in a magnetic tape or film caused by inadequate storage (warm temperatures) by which the signal from one layer is transferred to the next in a tightly wound roll. (2) Order, request made to the film laboratory to make the latent image (edge numbers) visible.

print up See *optical enlargement*.

Prix Gemeaux Les Prix Gemeaux; annual awards presentation by the Academy of Canadian Cinema and Television to honor excellence in the French language television productions of Canada. See also

Academy of Canadian Cinema and Television and *Gemini Awards, Genie Awards.*

probe lens   A lens system and accessories that includes a long lens barrel to enable the video or film camera operator to get over, under, or through objects being studied and photographed.

processing   A series of operations carried out in the motion picture laboratory to develop, rinse, fix, wash, and dry, etc. the exposed film to produce the negative or positive image. See also *development.*

process photography   Also called **process shot.** Photographic technique combining actual (live action) photography with the projected background of an illuminated set or design. See *traveling matte.*

process projection   See *back projection.*

producer   A person who organizes—often creates—and supervises a production, show, or series from inception (script selection and fund raising) to distribution and beyond. He/she is in charge of budgeting, the selection and hiring of production personnel and players (with the director), the overall production, distribution arrangements and often its financial outcome.

production   The actual making of a show, program, or film, involving all details, from preparation to final broadcast or screening.

production assistant (PA)   Also called **gofer** or **gopher.** A beginner or entry-level production crew member.

production credits   See *credits* and *title.*

production crew   See *crew.*

production designer   Art department head responsible for the design, setting, and the entire look, feel, and atmosphere of a program or film production. See also *art director.*

production executive   Managerial position at a production company, network, or station; the producer, director, story editor, supervisor, or other member of management.

production manager (PM)   Also called **unit production manager.** Person in charge of coordinating television or film production, including the scheduling of personnel and equipment, transportation, studio booking, and local arrangements.

production schedule   The listing of the day-to-day, hour-to-hour, and minute-to-minute activities of a program, show, or film. See also *log.*

production stills Photographs and/or slide pictures taken during actual production activities used for promotion and advertising purposes, or as aids.

professionalism A status of high standard in an activity and occupation marked by style, quality, and good taste. The mark of true professionalism starts with being on time.

professional quality See *broadcast quality.*

program controller Program manager. See also *controller of programmes.*

program length See *running time.*

programming/program department See *traffic department.*

progressive scan Full-frame video at the field rate. All scan lines are performed in a progressive sweep.

projection The process of presentation of the motion picture film on the screen for viewing. See *projector.*

projection axis Center of the projected picture from the projector to the center of the image on the screen.

projection booth Small room or enclosure, usually soundproofed, with one or more projection windows, housing the projector(s).

projection distance Also called **throw.** The distance between the lens of the projector and the screen.

projectionist Technician operating the projector.

projection print The film print used in projection instead of the original.

projection rate See *sound speed* and *silent speed.*

projection sync See *advance.*

projection throw See *projection distance.*

projector Equipment for viewing motion picture films by throwing (projecting) the images on the screen. Projectors may be for silent or for sound films, and are formatted for film sizes from 8mm to 70mm. Sound projectors are equipped with an optical and/or magnetic sound head.

promotion Also called **promo.** Promotional announcement; Broadcast station or movie theater publicizing its own upcoming programs to build audiences.

prompter Mechanical device attached usually to the camera, displaying words of the script to assist (prompt) the memory of actors, performers, announcers; or displaying the copy in continuity to be read on the air.

pronunciation check list　A list of hard-to-pronounce and/or foreign names and words to assist the announcer, narrator, or newscaster before and/or during a program or newscast.

propagation　Signals traveling outward, away from the transmitting antenna.

property release　A legal agreement, where required, by which the studio or production house receives the right to use the property's (house, yard, land,) picture in full or in part for production and/or promotion and publicity purposes. See also *model release.*

prop master　Also called **prop man/woman.** Property master; the person responsible for arranging and handling properties to be used on a show, program, or film.

props (properties)　All movable, physical objects used on the set in a program, show, or film, such as decorations, furnishings, hand props, vases, and dishes, etc. See also *hand props.*

protection copy　See *back-up copy.*

PRS　The Performing Right Society, Ltd. (GB)

PRTV　Public relations television.

PRU　ITU country code for Peru.

PRYC　Agencia Noticiosa Prensa Radio y Cine (Chile).

PS　Press Servis (Yugoslavia).

PSA　Public service announcement.

psi　Pound(s) per square inch.

PSN　Private satellite network(s).

P/S/N　PAL/SECAM/NTSC.

PST　Pacific Standard Time (USA).

PT　Prime time.

PTAR　Prime Time Access Rule.

PTI　Press Trust of India.

PTR　ITU country code for Puerto Rico.

PTT　Postal Telegraph and Telephone Authority.

PTV　(1) Public television. (2) Pay TV; pay television.

public domain (PD)   In the public domain. Material not protected or no longer copyrighted; i.e. available for performance without permission or remuneration.

public service announcement (PSA)   Announcements of public interest, in the service of the public, for which the broadcast station does not charge or receive payment.

Pulitzer Prizes, The   Annual awards administered by the Graduate School of Journalism, Columbia University, New York (USA). Honors are given in fourteen categories in Journalism, six categories in Letters, and one in Music; three fellowships are also awarded. The Graduate School of Journalism was endowed in 1903 by Joseph Pulitzer (1847–1911), a Hungarian-born American journalist and newspaper publisher, and the founder of the Pulitzer Prizes. The Graduate School of Journalism opened in 1912, after his death, and the Pulitzer Scholarships were instituted under his bequest in 1917.

pull   Generic term for *selection*.

pull down   The movement of the film, pulled frame by frame, in the camera or projector. See also *claw*.

pull focus   See *follow focus*.

pull stop   See *stop pull*.

pulse code modulation (PCM)   A type of modulation used in radar and satellite communication, where the carrier wave is broken up into pulses of varying lengths.

pulse sync (sync pulse)   Process in double system sound recording whereby the separate picture and magnetic tape recording is synchronized by generating a series of corresponding pulses on the sound tape. These pulses, when transferred onto a magnetic film, will fall in place with the sprocket holes to provide perfect synchronization.

push   Push processing; indication to force film development. See *forced development*.

pylon   A triangular, often towering structure erected for support, aerial, etc.

QAT ITU country code for Qatar.

QSL Confirmation of reception.

Q-TEL Qatar Public Telecommunications Corporation.

quad eight (quad 8) Five rows of sprocket holes alongside the 35mm film to render four strips of Super 8 films when slit. Used in laboratories making a large number of copies.

quadraphonic Sound reproduction system using four channels. See also *four channel.*

quadruplex head Quad head; four recording heads in a quadruplex video tape recorder.

quadruplex recording Also called **transverse scanning.** A 2-inch videotape recorder that makes the recording on four sections by employing four recording heads. Now obsolete. See also *helical scan.*

quality of sound A relative (and somewhat subjective) measure of sound (voice, music, effects) that is free of or has minimal faults in production. Quality of sound is achieved through proper casting, equipment selection and placement, and correct balance and frequency response in recording and reproduction.

1/4 inch Magnetic tape width standard, equals 6.35mm.

quarters Four times thirteen weeks, breakdown for programming television series. See *13-weeks.*

quartz Natural crystalline silica that exhibits a degree of piezoelectric effect.

quartz-iodine bulb A highly efficient incandescent light bulb that emits a very bright light.

quartz king A compact, high-intensity light with a quartz-iodine bulb.

QUASAR Quasi Stellar Radio Source; extragalactic source of high energy electromagnetic radiation (bright blue and ultraviolet light and radio waves).

quick study An actor or performer who is able to memorize lines and action in a short time. See also *understudy*.

quiz program A television question-answer program principally, derivative of old radio programs. Preparation requires good planning, a colorful setting, and careful selection of personnel and guests to enhance the program's visual appeal. See also *audience participation*.

QVC Quality Value Channel cable television network (USA).

R

R (1) Radio. (2) Film rating indicating that audiences under the age of 17 are restricted.

RA Radio Authority (GB).

RAB Radio Advertising Bureau (USA).

RACE Research into Advanced Communications in Europe.

rack (1) Device constructed of frames with rows of spindles (upper and lower) to carry film in a developing machine. (2) Storage rack; metal frame structure with multiple shelves to store film or tape in cans and containers. (3) Metal cabinet with spoked mounting holes to place and secure electronic equipment.

racking (1) Changing lenses by rotating the lens turret (in old style cameras). (2) "Rack into focus"—operating the focusing knob and/or ring in the cameras.

rackover The process of shifting the viewfinder into position on some non-reflex film cameras. After focusing and framing, the mechanism is racked back (over).

RADAR Radio Detection and Ranging; a general term for a system, developed during World War II, employing microwaves for locating, identifying, guiding, and navigating moving objects, like aircraft, ships, automobiles, missiles or artificial satellites.

radiant energy Energy transmitted in the form of electromagnetic radiation.

radiating element The radiant element. See *radiator.*

radiation The emission of electromagnetic waves, or light.

radiator The radiating element; a part of the aerial from which electromagnetic waves are emitted.

radio Also called **wireless.** Sound broadcast; the transmission of electrical signals without wires by using electromagnetic radiation. It includes radio, radio telephone, radio telegraph, the sound portion of television, and radar. Radio was invented by Italian physicist Guglielmo Marconi (1874–1937), a 1909 Nobel Laureate. See also *experimental radio* and *receiver.*

RADIOBRAS   Radio Brazil External Service.

radio frequency (RF)   The complete range of frequencies used for transmission by electromagnetic waves. 10 kHz to 100,000 MHz.

Radio Marti   Radio service to Cuba broadcast by the U.S. Information Agency.

radio microphone   See *wireless mike.*

radio relay   See *relay station.*

radio source   Term for the obsolete *radio star;* electromagnetic radiation of radio frequencies outside the solar system.

radio star   Obsolete term. See *radio source.*

radio telegraph   Also called **radio telegraphy.** Wireless telegraph; the transmission of coded messages (Morse code) by radio.

radio telephone   Also called **radio telephony.** Wireless telephone; telephone system operated by radio. See also *8XK.*

radio tube   British term for *valve.*

radio waves   See *radio frequency.*

RAE   Radiodifusion Argentina al Exterior.

RAI   Radiotelevisione Italiana.

rail   Power rail.

rain cluster   A set of water sprinklers suspended and connected with hoses to simulate rain.

RAIS   Rossiskoje Agentstvo Intellektual'noj Sobstvennosti Pri Prisidente Rossiskoj Federacii (Russia).

RAJA   Radio Joint Audio Research (GB).

rake   To place or shift a set into position. See also *strike.*

random noise   Random occurrences of undesired signals.

range finder   (1) A viewfinder with either a split image or double image that can be brought together by the adjusting ring. Range finders may be coupled to lenses and are found only on specific mounts (like helicopter mounts). (2) An optical device to measure the distance from the object to the camera.

raster   Evenly spaced, horizontal, parallel lines in a pattern scanning the television receiver screen (cathode ray tube). See also *interlace* and *scanning.*

rate　The time charge for a commercial announcement or a program.

rate card　Detailed schedule of advertising charges published by a broadcast station.

rating　(1) Audience rating; estimated percentage of viewers selected from a sample number of television homes. (2) Film emulsion speed.

ratio　See (1) *aspect ratio.* (2) *shooting ratio.*

raw stock　(1) Blank, unused magnetic (sound or video) tape. (2) Motion picture film that has not been exposed and/or processed.

RB　(1) Radio Bremen (Germany). See *ARD.* (2) Ritzaus Bureau (Denmark).

RCA　(1) Radio Club of America. (2) Radio Corporation of America.

RCC　Radio common carrier.

RCEMA　Radio Committee of the Evangelical Missionary Alliance (GB).

RCI　Radio Canada International.

RCN　Radio Cadena Nacional (Colombia).

R-DAT　Rotary digital audio tape.

RDDR　The former Radio der Deutschen Demokratischen Republik. No longer applicable.

reaction shot　A photographic shot, usually a close-up, of a performer's, bystander's, or on-looker's reaction. See also *cut away, nod shot,* and *reverse angle shot.*

reader　(1) Person who reads assigned literary material for possible production. (2) A small film editing room apparatus for fast viewing and monitoring of a picture or magnetic soundtrack, where the film is rolled or pulled by means of rewinds.

real focus　See *focal point.*

rear screen projection　See *back projection.*

rear projection unit　A small, portable, self-contained back projector with cartridge loading used in confined areas for sales, promotion, or education and training.

rebroadcast　Also called **delayed broadcast.** (1) A program that is recorded over the air, then rebroadcast over the station's transmitter. (2) A network program broadcast later at a different time by an independent station.

receiver (RX)   A radio or television receiver set. An apparatus, the final link in a broadcast chain, that transforms electromagnetic signals into sound and/or light waves.

receiving bay   A platform, adjacent to studios and storage areas, where props, large sets, equipment, and supplies are loaded and received.

rechargeable battery   A heavy duty DC battery pack unit or belt (NiCad) that can be plugged into a regular wall outlet and recharged after each use. Batteries come in 8, 12, or 16 volts for normal speed, and in 32 volts for high-speed cameras.

reciprocity   Law of reciprocity; exposure equals the intensity of light multiplied by the time of exposure: E = IT.

recorder   See *tape recorder* and *videotape recorder.*

recording head   See *head/2.*

recording speed   (1) Standard cyclic motion of a phonograph record or disc: 45 rpm, 78 rpm, or 33-⅓ rpm. (2) Magnetic tape speed.

recording studio   See *sound studio.*

recording system   Audio recording; a system combining microphone(s), mixer, amplifier, monitoring unit, and recorder used in sound recording with or without pictures. See also *videotape recording.*

recording tape   See *magnetic tape.*

recordist   Sound technician in charge of all stage (floor) recordings. Called **floor mixer** in Great Britain.

red   Additive primary color.

red flag words   Words and expressions that require caution and special attention because they may lead to a libel law suit.

reduced aperture   See *Academy aperture.*

reduction print   Motion picture film printed (reduced) from a larger size onto a smaller gauge film in the laboratory, i.e. a 35mm film onto 16mm or 16mm onto Super 8. See also *blow-up.*

reel   (1) Film or tape spool made of metal or plastic. (2) The tape or film wound on a spool or core. (3) Designation used for length of a finished film, e.g. three reels.

reference black   Also called **television black.** Zero percent picture signal voltage; shadowed or unilluminated areas with a reflection value of less than three percent. See also *reference white.*

reference print　Approved release film print kept for reference purposes.

reference track　See *scratch track.*

reference white　Also called **television white.** A fully illuminated white object with a maximum reflection value of sixty percent (in some cases 77%). See also *reference black.*

reflected light　See *bounced light.*

reflected light meter　Exposure meter that measures the intensity of light reflected from the subject, either as a whole or in a narrow angle (spot meter). See also *incident light meter.*

reflected light ultraviolet photography　See *ultraviolet photography/1.*

reflected wave　See *sky wave.*

reflector　Reflective lightweight aluminum surface mounted on a stand and used as a supplementary light source to sun- or studio light. It can be directed effectively to brighten dark or shadowed areas.

reflector lamp　Incandescent projection lamp with a built-in rear reflector.

reflex camera　Also called **mirror reflex** or **reflex shutter.** A film camera equipped with a mirror-like shutter that obscures the film during the pull down and at the same time reflects the image received through the lens onto a ground glass in the viewer.

refraction　The deflection of light or sound waves as they pass through a medium of non-uniform density, like the ionosphere or water.

Region I, II, III　See *ITU region.*

regional channel　A radio station with a power allocation that does not exceed 5,000 watts (5 kW).

registration　Also called **picture registration.** The precise positioning of an image in the aperture of a camera, projector, or printer. It affects the color in television, and the picture steadiness in a film camera.

registration pins　See *pins.*

regulator　See *voltage regulator.*

rehearsal　Preparatory practices before a live broadcast, videotaping, or filming.

rejects　Film rejects. See also *out-takes, bad footage,* or *bad take.*

relative aperture   See *f-number.*

relay station   A station located away from the main transmitter, which receives signals and retransmits them to another distant point or station.

release   (1) Legal release (model or property release), license, permission. (2) Press or publicity release. (3) Release print. (4) Film or video release; film distribution to theaters, television, and videotape.

release print   Also called **release**. Motion picture composite print, after the approval of the final answer (sample) print, released and distributed for showing in public cinema theaters. In the case of television broadcasts, the motion picture film is usually transferred onto videotape before telecast.

release print leader   Film leader designed to carry projection information, i.e. identification, synchronization and other information.

remake   New, updated version of a previously produced (older) program or film.

rem jet   An anti-halation layer on the back of the film base that is removed in processing.

remote   Remote pick-up; remote broadcast originating from outside the studio. Called **OB (outside broadcast)** in Great Britain.

remote control   An equipment control (wireless or by cable) away from the principal unit.

remote switch   On/off switch on a long cable, connected to the camera, recorder, or other equipment.

remote van   A specially equipped truck or van used on location for remote (OB) pick-up, transmitting audio and/or video signals via land (telephone) lines, microwave links, or satellite to the master control. Called **OB van** in Great Britain. See also *camera car.*

repeat program   A program retransmitted after its first broadcast.

replay   See *playback.*

replay head   See *playback head.*

reportage   The matter of the report or report program organized in continuity.

reporter   Staff member of a broadcast station who gathers, writes, or photographs and/or announces news items. See also *free-lance* and *stringer.*

report sheet   See *camera report.*

reprint   Additional film prints made from the original (or internegative) for editing purposes.

reproducing stylus   See *needle.*

re-recording   The mixing and combination of several soundtracks into one. See *dubbing/2* or *mix/1.*

residual income   (1) Profit received from repeat transmission of a recorded or filmed program. (2) Remuneration to an actor, performer, composer, or writer for the re-broadcast of a program or commercial. See also *royalty.*

resolution   Picture resolution; degree of screen reproduction of an image in fine detail, expressed in number of lines per millimeter defined in the image.

resonant circuit   Electronic circuitry with both the capacitance and inductance. Supplied by energy from a transistor or thermionic valve, it is used in receivers due to its selective detection and in transmitters for generation oscillations of radio frequency.

re-take   (1) The re-recording or re-photographing of a scene found unsatisfactory (N.G.). (2) The re-recorded or re-shot portion of the scene itself.

reticulation   Wrinkles or fissures, non-correctable breakup of the film emulsion, resulting from wide temperature changes during processing.

REU   ITU country code for Reunion.

Reuters (RN)   The British Reuters Telegram Co. news wire agency, founded in 1851.

reverberation   Echo in audio; the repeated reflection of sound in a closed environment. See *echo.*

reversal   Reversal film; motion picture film that produces a positive image after exposure and processing. See also *negative.*

reversal print   Reversal intermediate; color duplicate print made by reversal process.

reverse action   Projection of the film from its end to the beginning, reversing the flow of action.

reverse-angle shot   A photographic shot of a performer's face from the approximate point of view of another performer or interviewer. See *over-the-shoulder shot* or *reaction shot.*

rewind   Fast winding of recording tape or film backward; the opposite of *fast forward*.

rewinds   A pair of manual- or motor-driven geared devices used to roll (wind) film on spools or cores.

rex   Abbreviation for reflex.

RF   (1) Radio Frequency. (2) Radio Française.

RFA   Radio Free Asia (USA).

RF amplifier   An amplifier that operates at radio frequency.

RFE   Radio Free Europe (USA).

RFF   Radio Forces Françaises.

RFI   (1) Radio France Internationale. (2) Radio Frequency Interference.

RFO   Radio Télévision Française d'Outre-Mer.

RG   (1) Raidio na Gaeltachta (Ireland). (2) Radio Grenada.

RGB   Red, green and blue light in (1) Components of NTSC, PAL, SECAM color video. (2) Additive color film process. See *primary colors*/1.

RH   Relative humidity.

rheostat   Device used in film cameras to control electrical current by various resistance. See *wild motor*.

ribbon mike   Ribbon microphone; a sensitive ribbon or velocity microphone with high fidelity reproduction of sound, used extensively in studio work.

riding gain   The control, increase of signal power or transmitted sound by the studio engineer.

rifle mike   See *shotgun mike*.

rig   Colloquial term for equipment.

rigging   The placement of studio lights on a set.

ringing   (1) See *halo effect*. (2) Objectionable picture artifact in low-quality video recordings, most notable on transition edges of high-contrast images (white title letters).

riser   A small, low platform for elevating objects, cameras, or performers.

RITA   Russian Information Telegraph Agency.

RL   Radio Liberty (USA).

RM   Radio Moçambique.

RMR   Radio Moscow Relay.

RN   (1) Radio Nederlands. (2) Radio Nigeria. (3) Reuters News Agency (GB).

RNE   Radio Nacional de España (Spain).

RNT   Radio Nacionale Tchadienne (Chad).

role   The acting part, character in a broadcast program or film.

roll   Film wound on a spool or core. See also *roll film, roll sound, roll tape.*

roll drum   Roller title. See *crawl.*

roller title   Television and/or film title and credits slowly rolling— always—upward on the screen. See *crawl.*

"Roll film"   (1) Television control room cue to start film projection. (2) Direction to start film camera before a take, prior to "action."

rollover   Rolling of the television picture due to incorrect settings on the receiver set.

"Roll sound"   Roll, start sound tape.

"Roll tape"   Cue to start recorder. See also *Speed/5.*

room tone   Sound of an environment recorded before or after a "take," as atmosphere or background for editing purposes.

ROS   (1) Run of station. (2) Run of schedule.

rostrum   A square-shaped, portable, compact platform with folding legs.

rotary converter   An AC-powered electric motor coupled mechanically to a DC generator used to convert alternating current (AC) into direct current (DC).

rotating disc scanner   See *Nipkow disc.*

rotator lens   Lens that can be rotated 360° to make a multi-image montage from the same subject. Images can be made to appear in motion, upside down, moving clockwise or counterclockwise, etc.

rotary erase head   See *erase head.*

ROU   ITU country code for Romania.

rough cut   First assembly of a film in editing following the scripted order. See also *fine cut.*

routine sheet   See *log.*

routing switcher   The desired method in complex audio/video systems of sending signals to and from various equipment. Modern systems incorporate a matrix hierarchy for easy selection and may be computer-assisted.

royalty   Payment to a writer, composer, producer, director, actor, or singer for the sale/production/performance of artistic work and subsequent reuse. See also *residual income.*

R.O.Y.G.B.I.V.   Colors of the spectrum: red, orange, yellow, green, blue, indigo, violet.

RP (RSP)   Rear projection, rear screen projection.

rpm   Revolutions per minute.

rps   Revolutions per second.

RQT1, RQT2   Request One and Request Two pay television cable networks (USA).

RRW   ITU country code for Rwanda.

RSS   Rashtriya Sambad Samiti (Nepal).

RTB   Radio Televisyen Brunei.

RTBE   Radio and Television Broadcast Engineers (USA).

RTBF   Radiodiffusion Télévision Belge de la Communauté Française (Belgium). See also *BRTN.*

RTC   Radiodiffusion Télévision Congolaise (Congo).

RTCA   Radio and Television Correspondents Association (USA).

RTD   Radiodiffusion-Télévision de Djibuti.

RTDG   The former Radio and Television Directors Guild, now the *Directors Guild of America-DGA.*

RTE   Radio Telefis Eireann (Ireland).

RTF   Radio Télévision Française, the former ORTF (France).

RTG   (1) Radiodiffusion Télévision Gabonaise (Gabon). (2) Radiodiffusion Télévision Guinéenne (Guinea).

RTI   Radiodiffusion Télévision Ivoirienne (Côte d'Ivoire).

RTL   Radio-Tele-Luxembourg.

RTM   (1) Radiodiffusion Télévision du Mali. (2) Radiodiffusion Télévision Marocaine (Morocco). (3) Radio-Télévision Malagasy. (4) Radio Television Malaysia.

RTNDA   Radio Television News Directors Association (USA).

RTP   Radiotelevisão Portuguesa (Portugal).

RTS   Radio Television Seychelles.

rumble filter   Also called **hum filter.** Sound filter that cuts out low frequencies. See also *scratch filter.*

rundown sheet   See *fact sheet.*

RTT   Radiodiffusion Télévision Togolaise (Togo).

RTVE   Radio Television Española (Spain).

running shot   See *dolly shot* and *trucking.*

running time   The duration of a broadcast program or film in minutes and seconds.

run-out   Tail end of a tape or film.

run-out switch   See *buckle switch.*

run-through   See *camera rehearsal, dry run.* See also *rehearsal.*

run-up   Run-up time; in videotape recording a time period of 10–15 seconds is allowed for the tape to reach its correct operating speed.

RUS   ITU country code for Russia.

rushes   Rush print. See *dailies.*

RX   Receiver.

s   Second; 1/60th of a minute.

S   (1) ITU country code for Sweden. (2) Screen.

SABC   South African Broadcasting Corporation.

SACI   Brazilian award, somewhat like a combined Oscar and Emmy, given for cinematic and/or stage excellence in various categories.

SACEM   Societé des Auteurs, Compositeurs et Editeurs de Musique (France).

SACM   Sociedad de Autores y Compositores de Musicas (Mexico).

SAD   Second assistant director.

safe area   See *essential area.*

safe light   A type of light that does not affect film emulsion; used in processing laboratories and darkrooms.

safety film   Also called **non-flam film** or **polyester base film.** Safety base film; film that has a base with a slow-burning rate. See also *acetate/2.*

SAG   Screen Actors Guild (USA).

sales department   The station's principal link with the business community and advertisers, headed by the sales manager or director of sales. The sales department usually works in cooperation with an advertising agency or the client's representative. Sales may be national or local.

sample print   Rush film print with the general characteristics of a scene, but with no final grading corrections.

SANA   Syrian Arab News Agency.

San Francisco International Film Festival/Golden Gate Award Competition   Annual international film festival held in San Francisco, California, presenting the Golden Gate Awards for various lengths and categories (USA).

São Paulo International Short Film Festival   State government-sponsored international festival for short films up to 35 minutes in length, held in São Paulo (Brazil).

SAP   Secondary audio program.

SAPA   South African Press Association.

SAR   Society of Authors' Representatives (USA).

SARDC   Société des Auteurs, Recherchistes, Documentalistes et Compositeurs (France).

SAT   Satellite tier.

TCOM   Satellite communication.

satellite communication—SATCOM   Long-distance communication system utilizing orbiting artificial earth satellites in geosynchronous orbit 22,300 miles (appr. 35,680 km.) above the equator, without the limitations of submarine cable or the high-frequency (HF) carrier system. It provides global communication coverage with no bandwidth limitations, and can be used across land or sea. HF signals are bounced to and from ground antennae. Services provided by the satellite communication system include telegraph/telex, telephone, data, facsimile, monochrome or color television in the four ocean regions—East and West Atlantic, Pacific, and Indian Oceans. Participating nations are members of Intelsat. See *Intelsat*. See also *active satellite* and *passive satellite*.

satellite news gathering (SNG)   Broadcast news relayed by satellite using a highly mobile earth station that sends the signals to a satellite which then downlinks them to the receiving antenna.

satellite tier (SAT)   Cable television designation for satellite cable services and programmed channels from 14 and above. Does not include pay television.

Saticon   Trade name for high-quality, relatively low cost television camera tubes offering high resolution, good light handling characteristics, clean pictures, and excellent sensitivity. See also *Image Orthicon, Plumbicon,* and *Vidicon*.

saturation   Color saturation; the full rich purity of the color picture with the least admixture of white.

S-band   Frequency band of 2,535–2,655 MHz.

SBC   (1) Singapore Broadcasting Corporation. (2) Swiss Broadcasting Corporation.

SBCA   Satellite Broadcasting and Communications Association (USA).

SBE Society of Broadcast Engineers (USA).

SBS (1) Somali Broadcasting Service. (2) Spanish (Hispanic) Broadcasting System (USA). (3) Swaziland Broadcasting Service.

SC The Sports Channel cable television network (USA).

scale The minimum pay for actors and/or crew for their part in a broadcast program or film.

scanning Scanning beam; (1) Horizontal line sweep; the high-speed movement of the electron beam in a television picture tube, traveling from left to right in a sequence of lines from top to bottom, thus reproducing the transmitted image. See also *interlace* and *raster*. (2) A narrow slit of parallel light rays scanning the optical soundtrack in motion picture projection.

scanning area Television picture area reproduced on the control room monitor, but reduced during transmission on the home receiver screen. Similar to the *essential area* or *safe area* in film.

scenario The complete written version of a screenplay with details of scenes, characters, and appearances. The name applies usually to a film or television film, rather than to a teleplay.

scene A continuous portion of a teleplay or a film script intended to be played/acted and photographed in one set up.

scenery (1) Scenic objects, painted flats, etc., used to help to set the locale for a television program, show or film. (2) Spectacular panoramic, nature, location shots.

scene slating See *slating*.

schedule See *log*.

school broadcast Radio and/or television broadcast directed to schools and curriculum and containing directed activities, talks, debates, plays, conversations, and events (various sports and games). School broadcasts may be transmitted by educational stations, or by commercial stations at a designated time. See also *educational broadcasting/film*.

SCI-FI The Science Fiction Channel cable television network (USA).

SCN ITU country code for St. Kitts and Nevis.

scoop Appropriately shaped 500 or 1,000 watt television or film studio floodlight.

scope Abbreviation for (1) Oscilloscope or waveform monitor for video signals. (2) Wide-screen film system. See *Cinemascope*.

score The music used for a program, show, film, or commercial.

SCPC Single channel per carrier (audio).

scraper A small, toothed device in the film splicer used to remove the emulsion layer from the film base to ensure good splice cohesion.

scratch (1) Damaged groove on a disc. (2) Marks or scored lines on the film emulsion surface caused by dust or hard particles in uncleaned equipment.

scratch filter Sound filter that cuts out high frequencies and hiss (record surface noise). See also *rumble filter.*

scratch print See *slash print.*

scratch track Also called **reference track.** Soundtrack used solely for post-recording and editing purposes. See also *guide track.*

screen (1) TV screen; television picture (display) tube in a receiver. (2) A flat or curved light-reflecting surface upon which slides or motion picture film are projected. (3) See *barn door.*

screen brightness Also called **screen luminance.** The intensity of the light flux of the projector on the screen. It is affected by the screen surface, the lens, and the intensity of the light source.

screen credit See *credits.*

screening An advance showing or viewing of a broadcast program, film, or commercial.

screen luminance See *screen brightness.*

screenplay The complete written material for a broadcast show or motion picture film. In television it is often called a **teleplay.**

screen ratio See *aspect ratio.*

screen test A filmed test to evaluate a performer or actor for a particular part in a film.

screenwriter A person who writes film scripts, teleplays or treatments for motion pictures or television films. See also *Writers Guild of America.*

scrim A diffuser made of translucent material (gauze, silk, wire mesh, etc.) used to diffuse or reduce the intensity of light. See also *butterfly* and *gauze.*

script The written form of a radio or television program, film, or announcement with all audio and/or video material, including some directions. The text of a play used by the production personnel. Synonym for *continuity.* See also *shooting script.*

scripted order See *assembly.*

script or **continuity "girl"** Old form for **script supervisor.**

script supervisor Also called **continuity "girl"** (outdated). The director's assistant who keeps a record of all preparations, changes, shots, takes, etc., in connection with the script.

scriptwriter Staff or free-lance writer for continuity, treatment, screenplay, script, or teleplay.

SDA Schweizerische Depeschenagentur Agence Telegraphic Swiss—ATS (Switzerland).

SDG Screen Directors Guild. See *Directors Guild of America.*

SDN ITU country code for Sudan.

SDR Süddeutsche Rundfunk (Germany). See *ARD.*

SDTV Standard definition television.

season See *TV season.*

sec. Second.

SECAM Sequentiel Couleur à Memoire. See also *NTSC* and *PAL.*

secondary affiliate Broadcast station affiliated with more than one network. See also *primary affiliate.*

secondary audio program (SAP) A separate audio channel used for a second language, bilingual broadcast.

secondary movement Movement of the camera. See also *primary movement.*

second generation Also called **second generation dupe.** Second generation duplicate negative film. See also *first generation.*

second unit An additional camera crew to shoot various scenes, like establishing, mood, and scenery shots away from the actual set. A second unit consists of a cameraman/woman, camera assistants, essential transportation, and at times, and as budgets allow, a second unit director.

segue The sound transition of a musical number to another. A sound dissolve. See also *cross fade.*

selectivity The ability of a receiver to select a particular frequency signal from other nearby frequencies.

self-blimped camera A modern sound camera of controlled acoustical design. It has minimal camera noise which will not interfere with the microphone's sound pick-up.

Selsyn motor   Brand name for a self-synchronized motor in the film camera that provides interlock when coupled with the drive motor. See *interlock*.

semiconductor   The opposite of normal electrical conductors, i.e. its resistance decreases with rising temperature and the presence of impurities. Used in transistors and diodes.

semi-scripted show   A program with only partial indication of dialogue and/or action.

SEN   ITU country code for Senegal.

senior   Colloquial for a 5,000 watt quartz or incandescent studio spotlight with a Fresnel lens. See also *junior*.

sensitivity   See *emulsion speed*.

Sensitometer   Instrument that measures the sensitivity of a film or photographic plate.

separation light   See *backlight*. See also *lighting*.

separation master   Color separation master; red, blue and green components of an original negative prepared on separate strips with corresponding color filters on a panchromatic B&W track. A color intermediate negative or interdupe is then prepared with color filters.

separation negative   Color separation negative; individual strips of film, or single strips with successive frames, on which the component colors have been recorded separately as B&W images. In three-color processes red, green, and blue are recorded, and in two-color processes the blue-green and red-orange components are recorded.

sepmag   Film with a separate magnetic sound track.

sepopt   Film with a separate optical sound track.

sequel   (1) The continuation of a story or episode after a pause. (2) Another production of a film or the continuation (one week) later in television.

sequence   Incident in a film or television story recorded consecutively; equivalent to a scene in a play or a chapter in a book.

Sequentiel Couleur à Memoire (SECAM)   Color television standard developed in France that operates on a 625-line/50-field and is used in France, in most of the French-speaking world, the Middle East, parts of Africa, the Commonwealth of Independent States, and in some countries in East-Central Europe. See also *PAL, NTSC,* and Appendix E: Television Systems Worldwide.

series Serial; thematically grouped programs broadcast or screened over a number of quarters or weeks.

Service to Children Award See *National Association of Broadcasters.*

servo focus system Servo control; smooth, electrical zoom drive and focus control system on a zoom lens.

SES Societé Européenne des Satellites (Luxembourg).

set Setting; the context or arrangement of scenery and props in a scene. The place of action.

set designer The member or head of the art department who designs scenery—the entire setting of a television program or film. See also *production designer.*

set light See *background light.*

set-up The studio with performers, lights, sets, props, cameras, microphones, etc., ready for a take.

7 ½ ips Recording tape speed standard, equals 19.05 cm/s.

7-inch reel Diameter for magnetic tape reel, equals 17.7 cm/s.

7.62m 8mm film standard, equals 25 feet; now obsolete.

17.7 cm Metric measure for a 7-inch reel.

70mm film Film gauge application in wide-screen cinematography, in a ratio of 2.2:1. It is used as camera film in Russia and as prints in the USA and Western Europe based on a 65mm original, its width is 70mm, with 12.8 frames per foot.

78 rpm 78 revolutions per minute phonograph record speed. Now obsolete.

75mm lens Three-inch lens.

76.2 cm/s Outdated magnetic tape speed standard (used for mastering purposes), equals 30 ips.

72°F (22.2°C) Temperature at 45% relative humidity that is the ideal storage condition for tape or film. See *vault.*

SEY ITU country code for Seychelles.

SFB The former Sender Freies Berlin. (Germany). See *ARD.*

SFX (1) Sound effects. (2) Special effects.

SH Specified hours.

shading (1) The regulation of color. (2) The adjustment of black and white level. (3) The control of picture contrast.

shadow mask tube A three-gun single-screen television display tube used in color receivers.

shared screen See *split screen*.

sharpness See *focus*/2.

SHF Super high frequency.

shielded cable Coaxial cable protected (shielded) against external interference.

shipping case Metal, plastic, or fiber case that holds tape or film reels for convenient mailing and shipping.

SHN ITU country code for St. Helena.

SHO Showtime pay television network (USA).

shooting The photographic action with a camera.

shooting call Call time for actors and personnel involved in the actual shooting. See also *call*/1 and *crew call*.

shooting ratio Quantitative relation between the recorded tape or the composed film footage and the final edited version. The ratio varies greatly depending on the type of program shot: news type films—appr. 3:1; documentaries—4:1, 6:1 or more; feature films—very high; commercials—extremely high.
   The shooting ratio also depends greatly on good organization and the budgets allowed.

shooting schedule A list or timetable based on the script, arranged in groups of actions and scenes, with details of specific requirements of actors, locations, props, effects, transport, etc.

shooting script The final, detailed, written content of a television program or motion picture film, broken down into numbered scenes. See *script*.

short film Short subject in which the type and the content determines the length, usually not exceeding one or two reels.

short focal-length lens Wide-angle lens. See also *long focal-length lens*.

short pitch See *pitch*/2 and 2/a.

short subject See *short film*.

shortwave (SW) Electromagnetic wave of 6–12 MHz, 60m or less.

shorty   Called **baby legs** in Great Britain.

shot   (1) The single continuous take by the camera in one set-up. See also *take*/3. (2) The piece of film exposed by the single take (single run).

shot box   Control unit contained in a box for remote camera operations. Usually mounted on a camera panning bar, it can operate a zoom and gives (limited) panning and tilting. Some offer pre-set positions for zoom and focus of specialized lenses.

shotgun mike   Also called **gun mike** or **rifle mike.** Highly directional microphone that can be aimed toward the sound source like a shotgun or rifle. See also *omni-directional mike.*

shot listing   Film library classification of shots of released films; sometimes also includes out takes for stock footage.

shotmaker   Colloquial for a pickup truck-like vehicle with camera platforms mounted both in front and back. See also *camera car* and *camera truck.*

shot sheet   Also called **shot list, camera card,** or **crib card.** Separate list of television camera shots to aid the camera operator.

shoulder pack   Battery pack. See *battery.*

shoulder pod   Shoulder brace; portable camera support resting on the camera operator's shoulder to facilitate handheld operation. See also *body brace.*

shoulder shot   See *medium close-up (MCU).*

show print/copy   The finished (final) film print for special screening.

shrinkage   Film shrinkage; a reduction of film in both length and width due to processing and high-temperature drying, or as a result of long storage.

shutter   A mechanical rotating device in the film camera with an opening that can blank out the light reaching the film. Shutter openings either are fixed or are variable (adjustable) to a large degree with a minimum and a maximum opening. Shutters may be single-bladed (non-reflex cameras), two-bladed, or three-bladed. Some shutters rotate forward, some backward. High-speed cameras usually have electro-optical shutters. Modern cameras may be equipped with microprocessor-controlled silicon mirror shutters. See also *variable shutter.*

shutter opening   See *shutter.*

shutter speed   The exposure time of the shutter in motion at a standard camera speed. 1/48th of a second for 24 fps, and 1/50th of a second for 25/fps.

shuttle   Intermittent movement of the film in a camera or printer affected by claws.

SI Units System   Système Internationale d'Unités.

SICO   Servizio Informazione Chiesa Orientale (Vatican City State).

sidebands   The band of frequencies located at either side of the carrier wave. The width of the sideband is equal to the highest modulation frequency. A result of amplitude modulation (AM).

sidelight   Studio light, usually an addition to fill light, coming and directed from the side.

signal   (1) Electrical impulse conveying sound and visual information in a communication (transmission) system. Can also be electronic pulses or data generated by computer systems, which represent other forms of audio, video or graphic information. (2) See *cue/1* and *hand signals.*

signal-to-noise ratio (S/N)   The ratio of the signal transmitted to the ratio of noise in a transmitter and receiver system.

signatory   Networks, broadcast stations, studios and production companies who sign and are bound by contracts and make arrangements with respective unions and guilds.

signature   A characteristic audio (musical theme) or visual symbol of a particular program or commercial.

sign-off   The closing of the day's broadcast program usually followed by the national anthem(s).

sign-on   The starting announcement of a daily broadcast program giving the station's ID.

silent   A film or an insert with no sound. See *MOS* and *mute.*

silent frame   See *full aperture.*

silent speed   Camera and projection speed of film frames per second: 16 fps or 18 fps. In the silent era, film speeds varied between 12 to 20 fps. See also *sound speed.*

silhouette   Outline, shadow. See *cut-out/1.*

silk   White fabric diffuser on a frame attachment.

silver halide   See *halide.*

simulcast   A broadcast program aired simultaneously both on AM or FM radio, or on radio and television at the same time.

SIN   Servicio Iberoamericana de Noticias.

single-concept film   Low-budget film, mostly educational in nature, concerned with a single concept, such as numbers, trees, nature, training, or how-to subjects. See also *open-ended film*.

single 8mm film   See *8mm*.

single frame motor   Camera motor, used in single-frame shooting, that moves the film one frame at a time.

single frame shooting (single frame exposure)   A method of film exposure, used in animation and time-lapse cinematography, by which a single frame is exposed at a set interval.

single perforation (single perf.)   Motion picture film stock that has been perforated with sprocket holes on one edge only in the manufacturing process. See also *double perforation*.

single printer light   See *one-light print*.

single system   (1) Single system sound; the simultaneous recording of sound and picture on the same film, used in the past mostly in newsreel filming. The sound recording system may be optical or magnetic. See also *double system*. (2) Videotape recording which is a single system by design.

SIP   Swenska Internationella Pressbyran (Sweden).

SITA   Societé Internationale de Télécommunications Aeronautiques.

sitcom   Situation comedy.

16mm film   Small format, negative-positive motion picture film stock in width of 16mm, containing 40 frames per foot, with single or double perforations on the edges. It may have an optical or magnetic soundtrack.

60-field   See *525-line/60-field*.

6.35mm   Magnetic tape width standard, equals 1/4 inch.

61m   Standard film spool length, equals 200 feet.

625-line/50-field   Continental-European television standard. See also *525-line/60-field*.

65mm   A wide-screen cinematography camera original (negative) film stock, using 70mm duplication and release prints. It is 65mm in width with 12.8 frames per foot.

Sky Channel   Satellite channel, transmitting to Europe since 1982 and to Great Britain since 1984.

sky cloth   See *cyclorama*.

sky filter   Camera filter used in monochrome (B&W) photography to darken the blue sky. It does not affect other parts of the scene.

skylight   A 5,000- or 10,000-watt light bulb set used to illuminate a large (background) area.

sky pan   5,000-watt base and fill light with tungsten halogen quartz lamp, providing high-level illumination over a large area.

Sky Television   British DBS system, now part of British Sky Broadcasting—BSkyB (GB).

sky wave   Ionospheric wave; electromagnetic wave produced by transmitting aerials, directed to and reflected from the sky toward receiving aerials (MF, HF). See also *direct wave, ground wave* and *microwave*.

SL   Sign language; television viewing symbol indicating that the program is interpreted in sign language with the interpreter on-screen.

slant track   See *helical scan*.

slash print   Scratch print; a quickly assembled print made from the negative grading or scene-by-scene correction. Used as a guide.

slate   An information board, usually combined with a clap board, held in front of the camera before shooting, giving the title, the director's name, cinematographer's name, scene number, take, and date. New electronic systems generate a visual slate and a countdown audio tone.

slating   Cuing of the television or film program by recording/filming the slate board with the appropriate production information on it.

slave   A unit that is controlled by another in a two-unit operation.

SLBS   Sierra Leone Broadcasting Service.

sleeper   Colloquial term for a late blooming, but successful feature film.

sleeve   See *focusing rings*.

slide   Transparent, reversal (positive) still picture, usually mounted in glass plates. Sizes include 2¼ × 2¼-inch (6 × 6 cm) standard size; 3 × 4-inch (7.6 × 10.1 cm) for back projection; and 4 × 5-inch (10.1 × 12.7 cm) for station ID, titles, or inserts. Old style.

slitting   Method by which a film roll is split in half, or narrower width.

SLM   ITU country code for Solomon Islands.

sloat   A curtain track.

slo-mo   Colloquial abbreviation for *slow motion*.

slot   See *time, time slot*.

slotting   Inserting.

slow lens   Camera lens with a small aperture (high f-stop) for highly illuminated areas. See also *fast lens*.

slow motion   Slow motion effect; (1) In videotape recording: the images recorded with a camera/recorder fitted with technology that allows 90 fps versus the usual 30 or 25 fps. Also called **super slow motion.** (2) In motion pictures: the film run in the camera faster than normal speed. When projected at the standard rate, the motion will appear slower than normal. See also *fast motion*.

slow speed film   Motion picture film with a low ISO (ASA) or DIN rating.

SLP   Super long-play (VHS 6-hour speed).

SLR   Single-lens reflex (camera).

slug   (1) A short title indicating the content of a story in news copy. (2) A strip of leader replacing a missing or damaged film piece in the workprint.

SLV   ITU country code for El Salvador.

SM   Station manager.

SMA   ITU country code for American Samoa.

small film   See *substandard*.

small format   (1) Small size, portable video camera and recorder. (2) Narrow width videotape, 1 inch, 1/2 inch, or 8mm. (3) Narrow gauge film.

"Smart TV"   Colloquial for *digital television*.

SMATV   Satellite master antenna television.

SMO   ITU country code for Western Samoa.

SMPE   Society of Motion Picture Engineers, the founding name of the association—now *SMPTE*.

SMPTE   Society of Motion Picture and Television Engineers (USA).

S/N   Signal-to-noise ratio.

snap-on battery   Onboard battery; battery or battery pack attached to the camera. See *battery*.

SNBC   Sudan National Broadcasting Corporation.

sneak preview   Unadvertised public showing of a film to test audience reaction. Often followed by additional editing and adjustments.

SNG   (1) ITU country code for Singapore. (2) Satellite news gathering.

snoot   Also called **funnel.** Colloquial for a funnel shaped tube attachment affixed to a spotlight to reduce its light (spot) area.

snow   Electronic interference on the television picture tube that has the appearance of snow. See also *noise/2*.

SNV   Satellite news vehicle.

soap operas   Colloquial for daytime serial dramas. See also *telenovelas*.

SOCAN   Society of Composers, Authors, and Music Publishers of Canada.

Societé Européenne des Satellites (SES)   A private European organization operating the ASTRA satellites since 1988, with emphasis on DTH (direct-to-home) broadcasting services (Luxembourg).

Society of Film and Television Arts   A former society formed by the British Film Academy and the Guild of Television Producers and Directors. In 1957 it developed into The British Academy of Film and Television Arts (GB).
See *The British Academy of Film and Television Arts*.

Society of Motion Picture and Television Engineers (SMPTE)   A technical organization for professionals in the motion picture, television, electronic imaging, and related arts and sciences fields (USA). Founded in 1916 and named the Society of Motion Picture Engineers (SMPE), it incorporated television in 1950.
The Society provides a forum for equipment standardization, materials and practices and disseminates relevant technical information. It publishes the monthly *SMPTE Journal; News & Notes,* a monthly newsletter; and the annual *Progress Report and Directory of Members.*

SOF   Sound on film.

s/off   Sign off.

soft   (1) A magnetic film or a print with low contrast. (2) Soft light.

soft focus   A diffused effect where the subject appears in less than sharp focus. It may be achieved with a camera filter, or sometimes it is simply a mistake.

soft light   Studio light with a soft, diffused lighting effecting a smooth field. See also *hard light*.

solar cell   Battery; a device, usually a semiconductor, powering electric circuitry in satellites by converting sunlight into electrical energy.

solid state circuit   Electronic device, such as a semiconductor, transistor, or integrated circuit, with no moving parts, heated filaments, or gases.

solid state video camera   Smaller, lighter, stronger video camera using CCDs (charge-coupled devices) instead of vacuum tubes as the image sensor.

SOM   ITU country code for Somalia.

s/on   Sign on.

SONAR   Sound Navigation Ranging; an apparatus that transmits special sound waves and receives the reflected wave; used to locate underwater objects. The travel time of the sound wave indicates depth.

SONNA   Somali National News Agency (Somalia).

SOP   Standard operating procedure.

SOR   Servicio Oficial de Radiodifusion (Argentina).

SOS   (1) Sound on sound. (2) The radiotelegraphic international distress signal, a call for help.

SOT   Sound on tape.

sound   See *audio*.

sound advance   See *advance*.

sound amplifier   See *amplifier*.

sound baffle   See *baffle*/3.

sound barney   See *barney*.

sound blimp   See *blimp*.

sound booth   See *announcing booth*.

sound broadcast   See *radio*.

sound camera   See *self-blimped camera*.

sound console   See *console*.

sound drum   See *drum*/1.

sound effect library   See *effects library*.

sound effects (SFX)   Sounds, like street noise, steps, or animal sounds, added to dialogue, narration, or a musical score. Sound effects may be natural, prerecorded, created, or re-created. See also *optical effects* and *special effects*.

sound effects track   A separate sound track containing the sound effects.

sound engineer   Technical personnel who records sound and sound effects.

sound film   Motion picture film with accompanying sound portion—music, narration, dialogue, sound effects—on either an optical or a magnetic track along the film's edges. Partial sound on film was introduced by the Warner Brothers' film *The Jazz Singer* in 1927. The first full sound film (talkie) was *The Lights of New York* in 1928 (57 min. long), also a Warner Brothers production, directed by Bryan Foy.

sound film speed   See *sound speed.*

sound head   Tape recorder head used for recording, erasing, and replay.

sound lock   A chamber with quiet operating doors at studio entrances enclosing a soundproof area.

sound log   See *sound report.*

sound master positive   A low-contrast positive print derived from the original sound film or tape, used for preparing dupe negative.

sound mixer   See *console* and *mixer/2,3.*

sound mixing   See *dubbing.*

sound on film (SOF)   Single system filming.

sound on sound   A sound recording method where material previously recorded on one track may be rerecorded on another track, while new material is simultaneously added.

sound perspective   A feature whereby the picture and the accompanying sound must be in perspective, i.e. close shots (CU) must have close sounds and long shots (LS) must have faraway sounds.

sound projector   Film projector designed and equipped to project films with magnetic and/or optical sound.

sound quality   See *quality of sound.*

sound reader   See *reader/2.*

sound recorder   Sound recording channel; consists of the microphone(s), mixer, and recorder chain. This term also refers to the sound camera.

sound report (sound log)   A report filled during recording—similar to the camera report—indicating takes, dates, location, and speed and type of recording.

sound speed (sound film speed)   Standard speed of film at 25 frames per second for all sound films and television films in the 50-field systems. In 60-field television systems, the sound speed is 24 fps. See also *silent speed*.

sound stage   Acoustically and climatically controlled studio or building where sound recording and sound filming are done.

sound stock   See *magnetic film*.

sound stripe   See *magnetic stripe*.

sound studio   A soundproof studio used for announcing, recording, and mixing.

sound tape   See *magnetic tape*.

soundtrack   (1) Magnetic track of the tape containing the sound portion. (2) Magnetic tape or film with sound (music, dialogue, sound effects) to be matched in editing with the visual portion of the film. See also *optical track*.

sound velocity   The speed of sound waves traveling through air at the rate of 1,120 ft or 332 m per second (appr. 760 miles or 1,260 km per hour) at 0° Celsius.

soup   Colloquial for (1) Fog or haze on location. (2) Chemical solution in a film processing laboratory.

Sovcolor   Negative/positive integral tri-pack color film process based on the *Agfacolor* process; made in the former Soviet Union.

SP   (1) Standard play (VHS 2-hour speed). (2) Superior performance, as applied to Sony's *U-matic* and *Betacam* (SP Beta) videotape formats. (3) Servicio Nacional de Prensa (Colombia).

spacing   Silent, blank film pieces used in editing double system film to fill sections where the synchronized soundtrack is missing. See also *slug/2*.

spaghetti   See *buckle*.

SPDIF   Sony/Philips Digital Interface Format.

speaker   See *loudspeaker*.

speaker output   See *output*.

special effects (SFX)   (1) Electronic effects created for video. (2) Trick photography using miniatures, models, split screens, electrical and mechanical devices, or computers involving in-camera techniques, laboratory processes, or both, to achieve scenic or dramatic effects impossible or too costly to produce with actual, routine filming techniques. See also *optical effects* and *sound effects.*

special events   Broadcast programs of national or international importance covering items not regularly scheduled, like emergency operations or political and sporting events.

spectrum   Electromagnetic spectrum; the emitted energy of a radiant source as arranged in wavelengths.

speech channel   Also called **telephone channel.** A channel with frequency ranges from 250 to 3,400 Hz.

speed   (1) See *magnetic tape speed.* (2) Light-transmitting power of the lens, expressed in f-stops (f-numbers) or T-stops (T-numbers). (3) The rate of film movement in the camera, projector, or another equipment measured in feet per minute (fpm), centimeters per second (cm/s), or pictures per second (pps). (4) Film speed; see *emulsion speed.* (5) "Speed"—reply to "Roll tape" command in taping/filming, indicating that the sound recorder or video recorder is rolling at speed, ready for a take.

SPG   Screen Publicists Guild (USA).

spherical antenna   Multi-focus antenna.

spider box   Electrical connection box with several sockets for fast connection of cables and lights.

spider dolly   A versatile eight-wheel dolly that turns, twists, and tracks in any direction; it features a hydraulic lift for smooth action.

spillover   Broadcast program reaching audiences in neighboring countries, spilling over frontiers.

spindle   See *capstan.*

spinoff   Imitation of a successful broadcast program with a different, new angle.

splice   The point where two pieces of tape or film are joined. See *butt splice* or *lap splice.*

splicer   A mechanical device of various types used by the editor (film cutter) to join film or tape pieces. See *guillotine splicer.*

splicer tape   Transparent adhesive tape in different sizes, with or without sprocket holes, used in tape or film editing.

splicing   The process of joining film or tape pieces in editing.

splicing block Small metal block with a curved groove for magnetic tape editing. It usually contains two guide slots for diagonal (sound) and vertical (picture) editing.

splicing cement See *cement.*

splicing machine See *splicer.*

split field lens Also called **diopter lens.** A camera filter lens with one half blank, the other half being a close-up diopter. It allows subjects in the distance and foreground to be filmed in far focus.

split focus Focus adjustment somewhere between two or more widely separated objects to include them in the depth of field for sharp definition or dramatic value.

split reel Film spool (reel) with removable flanges for easy loading/unloading of film on core.

split screen Also called **shared screen.** Two or more separate pictures combined in an effect shot. In television the output of two (or more) cameras is combined. In film it is achieved by a laboratory process.

SPM ITU country code for St. Pierre and Miquelon.

sponsor The advertiser; a company, firm, or individual who pays for the film, broadcast time, or program in which products are advertised.

sponsored program An advertising program for which the station receives payment.

spool Tape or film reel with a flange.

sportscast Broadcast program covering sports events.

spot Short commercial advertisement, usually pre-recorded.

spotlight A studio light with focusing capability to provide key illumination.

spot meter Single-lens reflex type light meter for critical spot measurement.

spotting (1) Locating single words or sounds during the editing process. (2) Marking the dialogue on the soundtrack for foreign language captions. (3) Locating and identifying persons and personalities on location and at sporting events, community or political gatherings.

spread (1) Time spread, available time for stretching a program. (2) Time allotted for audience reaction, laughter, and applause during shows with live audiences. (3) The width and area coverage of a light.

spreader   See *triangle*.

sprocket   Sprocketed wheel; part of the film camera, projector, or editing equipment to align and transport film along its perforations.

sprocket holes   See *perforations*.

squawk box   Slang for CB radio or intercom speaker.

squeegee   Wiping blade or roller, used in continuous processing machines to wipe excess liquid off film surfaces. See also *air knife*.

squeeze   Anamorphic horizontal compression, "squeezing" in wide-screen film process.

SSR   (1) Saarlandischer Rundfunk (Germany). See *ARD*. (2) Sveriges Radio (Sweden).

SRA   Station Representatives Association (USA).

SRC   Societé Radio Canada-CBC.

SRG   Schweizerische Radio-und Fernsehgesellschaft-SSR (Switzerland).

SRL   ITU country code for Sierra Leone.

SRP   Standard and recommended practices.

SS   Scientific (non-commercial) satellite.

SSAD   Second second assistant director.

SSB   Single side band.

SSR   (1) Societé Suisse de Radiodiffusion et Télévision-SRG (Switzerland). (2) Societa Svizzera di Radiotelevisione (Switzerland).

STA   Slovenska tiskovna agencija (Slovenia).

stability   (1) Stable motion in a sound recorder. (2) Balance, steady operation of the television picture. (3) Steady film camera or projector.

stabilization   The fixing of the image in chemical solution in the film laboratory. See also *stability*.

stabilizer   (1) See *voltage regulator*. (2) See *Dyna lens* and *gyro head*. (3) See *fixing*.

stacked heads   See *in-line heads*.

stage   Generic term for studio. See also *sound stage*.

stage brace   Support for scenery or flats.

stage hands   See *floor men/women.*

stage left/right   Direction from the actor's point of view (from the stage) facing the audience. The opposite of camera left/right. See also *camera left/right.*

stage manager   Floor manager.

stage plan   Floor plan.

stagger through   Colloquial British term for *walk-through.*

staggered heads   Two half-track recording heads following each other on a tape path. First track creates left channel, second head the right channel. In physically edited magnetic tapes it created problems. Now obsolete.

standard candle   See *candela.*

standard 8mm film   See *8mm.*

standard lenses   Wide-angle, normal, and telephoto lenses in a three-camera turret (old system).

standards conversion   The change of television signals from one standard to another by electrical and optical techniques, from NTSC to PAL or SECAM and vice versa. See *field rate converter, line rate converter,* and *line store converter.*

stand by   (1) Performer, announcement, or substitute program held in reserve for emergency use. (2) Order given in the studio (on the set) to get ready for the starting cue.

stand-in   An extra who takes the place of a lead or important actor during the set-up. See also *double.*

standing wave   Also called **stationary wave.** A wave resulting from the simultaneous transmission of two similar wave motions in the opposite direction.

star   (1) Popular name for the leading actor/actress (USA). (2) Star (Muslim) News Agency (Pakistan).

star effect filter   Starburst filter; camera lens filter that gives the effect of one or several gleaming stars.

start mark   Marking on film and tape to indicate the synchronization (starting) point.

STAR TV   Satellite Television Asia Region (Hong Kong).

static   (1) Noise heard in the receiver as a result of atmospheric or man-made electrical disturbances. (2) Thin spread marks on film emulsion caused by the discharge of static electricity.

station identification   ID; Station break. See *break*/2.

station licensing   The authorization and allocation of services of a broadcast station and the franchising of cable television systems by the Federal Communications Commission/FCC (USA).

stationary orbit   See *synchronous orbit.*

stationary wave   See *standing wave.*

station promo   Station promotion; a series of short announcements aired by broadcast stations to promote an event, program, or series.

STD   (1) Standard tape. See also *LN*/2. (2) Abbreviation for *studio.*

Steadicam™   Trade name for a handheld camera support system that allows movement by the operator, like a walking follow shot, running up and down stairs, and panning, tilting, to ensure steady picture registration. See also *body brace.*

step-down transformer   Apparatus for reducing electrical (AC) voltage.

stepped lens   See *Fresnel lens.*

step printer   Film printing apparatus moving the negative and positive intermittently, step-by-step, and stopping each frame for the time of exposure. See also *contact printer* and *optical printer.*

step wedge   Test film with a series of exposures of increasing density, used to determine film stock characteristics in the laboratory.

stereo   Stereophonic sound; a high fidelity recording and reproduction process with dynamic range, providing binaural hearing by way of laterally placed microphones and two separate speakers.

stereoscopic cinema/television   Motion picture or television process where, in addition to the dimensions of width and height, a degree of measurable depth is also presented.

stereoscopic picture   Two separate pictures recorded so that when viewed by normal binocular (two-eyed) viewers, it appear as one, giving a measurable dimension of depth. See also *Vectograph.*

stereopticon   An improved magic lantern for projecting images.

sticking   See *burn-in.*

still   Still photograph; photographic print or slide.

still frame   See *freeze frame.*

STIMAD   Societé des Télécommunications Internationales de la Republique Malagache (Madagascar).

stimulated emission   See *laser* and *maser*.

stock   See *raw stock*.

stock footage   Stock shots, film footage of various events, historic or famous places kept in the film library for later use. Also a collection of frequently needed general shots of traffic, street scenes, crowds and the like.

stock footage library   See *film library*.

stop   See *f-stop* or *T-stop*.

stop bath   Processing solution to stop the chemical action in film development.

stop down   The reducing of the lens aperture.

stop frame   See *freeze frame* and *single-frame shooting*.

stop motion   Stop motion cinematography; exposure of the film in the camera one frame at the time—a basic requirement in animation. Also used to record very slow, time-consuming movements, like plant growth. See also *time lapse cinematography*.

stop pull   Pull stop; the action of altering the lens aperture (f-stop) during a continuous shot, done usually by a camera assistant. See also *pull focus* and *follow focus*.

stopping down   See *stop down*.

storage   See *vault*.

storage rack   See *rack*/2.

story board   A series of drawings or stills with accompanying text to indicate the outline of a story or scene, each change of action being represented by a picture. Mostly used for commercials or animation.

story editor   A writer who reviews and/or edits literary or dramatic material for television or film series.

STP   ITU country code for São Tome e Principe.

straight cut   Vertical cut used in film and tape editing to join film or tape pieces. See also *diagonal cut*.

stretch   Cue to slow down and pace the program (stall for time).

strike   (1) Take down, disassemble the set or parts of it. See also *rake*. (2) Discontinued work as a union bargaining technique.

stringer A reporter, camera operator, or correspondent (free-lance or on staff elsewhere), who does part-time work for a broadcast station or (news) organization. See also *free-lance*.

strip See *filmstrip*.

stripe See *magnetic stripe*.

strip light A series of interconnected lamps assembled in a strip to illuminate a cyclorama.

strobing Stroboscopic effect; a visual effect achieved by fast panning over static objects like railings, spoked wheels, or fences.

STT-FNB Suomen Tietotoimisto-Finska Notisbyran (Finland).

studio The acoustically controlled stage or workshop, where radio, television, and film programs are produced. Also a general term for a production facility complex with several single studios and related services.

studio camera A large camera mounted on a pedestal, movable dolly, or crane and used in a controlled studio environment. See also *field camera* and *portable camera*.

studio floor manager See *floor manager*.

studio monitor See *monitor/2*.

stunting Colloquial for broadcast program tactics employing abrupt changes, new series (often in mid-season), and heavy promotions designed to throw competing scheduled shows off-balance.

stunt man/woman A substitute performer hired by larger studios to take the place of a lead actor in dangerous situations, fights, and/or to perform specific acrobatic acts, like a car crash, jumps, etc., for the camera. See also *double*.

Preston Sturges Awards Annual award by the Directors Guild of America for both writing and direction. See also *Directors Guild of America*.

STV Subscription television (Europe).

stylus See *needle*.

subcarrier A special constant frequency signal within the complex encoded television signal, used as a reference for color information and sound.

subjective shot A moving camera shot that shows the scene as viewed through the eyes of the character. See also *camera helmet*.

subscriber A home connected to cable television or pay TV service.

substandard Substandard film; outdated film term for a narrow gauge motion picture film less than 35mm width, i.e. 16mm, Super 8mm, or 8mm. See *narrow gauge film*.

subtitle Translation of a foreign language dialogue and/or narration superimposed on a film or television image. See *caption*.

subtractive color process The basis of color film process in which transparent dyes filter out (subtract) the unwanted components from the white light spectrum. The subtractive primary colors are cyan, magenta and yellow. See also *additive color process*.

subwoofer A speaker that reinforces bass (low-frequency) sounds in the 10Hz to 40Hz range. See *woofer*.

SUI ITU country code for Switzerland.

SUNA Sudan News Agency.

Sundance Film Festival Utah-based annual festival for documentary, short, general subject, and American independent feature films. Honors include the Audience Award Grand Jury Prize, Filmmakers' Trophy, Waldo Salt Screenwriter Award, Cinematography Award, and Freedom of Expression Award (USA).

sun gun Compact, powerful, handheld light with a rechargeable battery, used mainly in news filming and electronic news gathering (ENG).

sunlight See *natural light*.

sunshade See *lens hood*.

Super Channel ITV satellite channel broadcasting BBC and ITV programs (GB).

Super 8 Super 8mm film; narrow gauge film of a better quality and larger frame size than the standard 8mm film it replaced. Used extensively in education, marketing, industrial training, and home movies both as an original camera film, and as a reduction print from 16mm originals. Super 8 has 72 frames to the foot.

super high frequency (SHF) 3,000–30,000 MHz with a wavelength of 1–10cm.

superimposition Often called **super.** The overlapping and blending of two images in such a way that each will show through with good definition. In television it is achieved electronically, in film by a laboratory process. See also *double exposure*.

superimposed title (1) Title and credits superimposed over scenery, action and/or a photographed background. (2) See *subtitle*.

Super-Panavision   See *Panavision/2.*

Superscope   Wide-screen film presentation technique producing copies anamorphically from 35mm negative, which are then projected through an anamorphic lens.

Super 16   Motion picture original, with appr. 24% larger frames than the regular 16mm film and with an aspect ratio of 1.66:1 (versus 1.33:1). Designed as a camera film for blow up of exact proportion to 35mm film and/or to wide-screen release prints.

super slow motion   See *slow motion/1.*

superstation   Independent station providing programs via satellite to a number of stations and cable systems.

supply reel   Reel with tape or film wound on it, feeding the take-up reel in a recorder, camera, or projector. See also *take-up reel.*

SUR   ITU country code for Surinam.

surround sound   See *four channel.*

sustaining program   A station-supported broadcast program for which it receives no remuneration.

SVN   ITU country code for Slovenia.

SW   Shortwave.

swaying   Slow, regular sideways shifting of a picture.

sweep   (1) See *scanning.* (2) A piece of curved scenery. (3) Colloquial term for panning.

sweetening   Colloquial for the mixing and combining of various sound sources, equalizing and balancing them for a composite sound (audio) track.

SWF   Südwestfunk (Germany). See *ARD.*

swish pan   A fast panning shot, achieved by swiftly panning the camera, thus creating a blur on the film. Used mostly as a transition.

switch   Change from one lens to another, from one angle to another, or from one camera to another. See also *cut/3* or *take/1.*

switcher   (1) Audio mixer. (2) See *vision mixer.*

SWR   Südwest Rundfunk (Germany).

SWZ   ITU country code for Swaziland.

Sydney Film Festival   International festival of feature, documentary and experimental films in 16mm, 35mm and 70mm formats (Australia).

sync generator   See *sync pulse generator.*

synchronization (sync)   The correct position of sound and picture in relation to each other; the coordinated match of audible and visible components in television and/or film. See also *lip sync.*

synchronizer   (1) Film to video: synchronizer to correct speed and eliminate rolling (horizontal bar) in the monitor or television set during filming. (2) Gang synchronizer: a device used in the editing room to measure film and to keep two or more pictures and the corresponding soundtracks in perfect synchronization. It is equipped with sprocketed rollers, frame and footage counters, a lock, and possibly a sound reader.

sync mark (synchronizing mark)   Mark made by the film editor on the picture and soundtrack to maintain continuous synchronization.

sync motor   See *governor motor.*

sync pulse   (1) See *pulse sync.* (2) See *sync pulse generator.*

sync pulse generator   Synchronizing pulse generator; the generator that produces the several synchronization pulses to keep the camera pick-up device in perfect synchronization with the display tube in the receiver/monitor.

Modern sync pulse generators are small and are built in to cameras, professional video recorders, and special effects mixing and switching systems. They can **gen-lock** to a common stable video signal to achieve absolute stable synchronization.

synchronous orbit   Also called **Clarke Belt, geosynchronous orbit,** or **geostationary orbit.** Stationary orbit; an artificial earth satellite whose orbit has a period of 24 hours in a corresponding altitude of 22,247 miles or 35,800 kilometers, parallel to the equator. Communication satellites in synchronous orbit are used to relay radio signals in long distance (widely separated) global transmission.

synchronous speed   Speed identification of camera, sound recorder and/or projector to ensure proper picture movement and corresponding sound pitch: 24 fps on 60Hz, and 25 fps on 50Hz power systems.

sync roll   Loss of synchronization and vertical roll of a television picture.

sync sound   Synchronized sound.

syndication   The sale of programs and/or series to individual, not interconnected broadcast stations. These programs exclude live presentations.

synopsis A detailed "blueprint" for a screenplay or teleplay, written in action sequences, but without full technical data or continuity. Carried out between writing the outline and the treatment.

synthetic processing Synthetic imagery; creating previously non-existing products in computer animation versus existing images. See also *image processing*.

Système Internationale d'Unités (SI) The international units based on the meter-kilogram-second system. Its basic units are: the meter (m), kilogram (kg), second (s), ampere (A), Kelvin (K), mole (mol), and candela (cd). The hertz (Hz), watt (W), volt (V), lumen (lm), and lux (lx) are among the derived units.

System 35 A combination film and television camera system in single- or multi-camera film production. A Vidicon camera tube or CCD sensor is attached to the blimp door of the film camera and its output is fed to an electronic viewfinder. Television output is relayed and switched at a mixing console for shot selection, and a VTR preview unit is attached for instant replay. See also *video assist*.

SYR ITU country code for Syria.

T Time.

table stand Also called **desk stand.** Small, adjustable microphone support placed on a table.

tachometer Measuring instrument graduated in frames per second (fps) to indicate the speed of a wild motor in the film camera.

tag Tag line; the last line of the copy, usually delivered with some emphasis.

tail The end part of a roll of tape or film.

tail leader Leader at the end of a tape or film for protection. It usually carries identification marks.

tail slate See *end slate*.

take (1) Television control room cue to cut or switch instantaneously from one camera to another ("take one," "take two"). (2) A single, continuously recorded sound portion. (3) A scene or part of a scene filmed in one continuous shot. See also *bad take* and *good take*.

"Take a level" See *voice test*.

"Take it away" Colloquial cue to start a remote (OB) broadcast program—"It's on the air."

take-up reel Reel or spool onto which magnetic tape or film is wound in a recorder, camera, or projector, fed by the supply reel. See also *supply reel*.

talent An actor or performer.

talkback One-way communication system from the control room to the studio. A special intercom system.

talkies Early name for sound films.

talk program Broadcast program consisting of panel discussions (talk panel), interviews, conversation, lectures, and/or telephone question-and-answer. The program may be live or pre-recorded. See also *interview program* and *panel/2*.

tally light Small red light on the top of a television camera indicating when that particular picture is being transmitted on the air. Called **cue light** in Great Britain.

Tampere International Short Film Festival Annual competitive festival for short fiction, documentary and animation films held in Tampere (Finland).

tank Developing solution container in the film laboratory.

TAP Tunis Afrique Presse.

tape See *magnetic tape*.

tape deck A tape recorder/playback unit designed for use in a built-in high fidelity music system. It usually consists of amplifiers for recording and pre-amplifiers for playback.

tape duplication Multiple copies of magnetic tapes made from the (master) original on recorders or duplicating machines. Special audio duplicators are equipped with at least two speeds: 7-1/2 ips and 15 ips. The latter for high speed operation.

tape editing Videotape editing; using an electronic device incorporating digitally generated numbers for easy (visual) identification of each frame.

tape guides Grooved metal posts on either side of a head assembly in a recorder that keep the tape in its proper track as it travels across the heads and through the entire tape path.

tape index counter A digital counter used to measure the portion of tape passing across the heads. See also *tape timer*.

tape joiner See *splicer* and *guillotine splicer*.

tape recorder Apparatus for electromagnetic recording and instantaneous reproduction of sound signals. See also *VCR* and *VTR*.

tape sizes See *magnetic tape width*.

tape speed See *magnetic tape speed*.

tape splicer See *splicer*.

tape splicing block See *splicing block*.

tape storage See *vault*.

tape timer A mechanical or electronic device that times the recording or playback of tapes in hours, minutes, and seconds.

tape transport The mechanical parts in a recorder, i.e. motors, reels, controls, etc., that facilitate the transport of the tape.

target (1) Light-sensitive part of the television camera pick-up tube on which the electron beam is scanned to form an image. (2) A circular-shaped small flag. (3) Audience for whom a specific (targeted) program or film is intended.

TA-SR Tlakova Agentura Slovaskej Republiky (Slovakia).

TASS The former Telegrafnoje Agentstvo Sovietskovo Soiusa. Now called **Telegrafnoje Agentstvo Suverenykh Stran** (Russia). See *ITAR-TASS*.

TBC Time-base corrector.

TBS (1) Tokyo Broadcasting System (Japan). (2) Turner Broadcasting System (USA).

TC (1) Telecine. (2) Technical coordinator.

TCA ITU country code for the Turks and Caicos Islands.

TCD ITU country code for Chad.

TCM Turner Classic Movies cable channel.

TCRB Television Code Review Board (USA).

TD Technical director.

TDC The Discovery Channel cable television network (USA).

TDF Télédiffusion de France.

TDM Teledifusão de Maçau.

tear Aberration of the television picture caused by various disturbances.

teaser (1)Black cloth hung on the set to shield the light from hitting the camera lens. (2) Advertising/promotional shorts for a program or film, cut dramatically and withholding major information, to arouse interest and curiosity.

technical director (TD) The person in charge of the technical crew and operations in the broadcast studio. In the USA he/she operates the video switch panel. See also *vision mixer*.

Technicolor Trade name for an American integral tri-pack color film process used in professional cinematography. Developed by Herbert Thomas Kalmus (1881–1963), American physicist.

Technirama Wide-screen film production technique, using a double-size frame on a 35mm negative, photographed anamorphically in a ratio of 1.5 to 1.

Techniscope   Production system for wide-screen film presentation employing half-size frame photographed without distortion on 35mm negative. Using anamorphic enlargement, full-frame 35 mm copies are printed for projection anamorphically.

TELAM   Telenoticiosa Americana (Argentina).

TELCOR   Instituto Nicaraguense de Telecomunicaciones y Correos (Nicaragua).

telecast   Broadcast television program.

telecaster   Television broadcaster.

telecine (TC)   (1) Television film and slide projection chain and the room where the equipment is placed. British term. See *film chain*. (2) Film made for television use.

telecom   (1) Telecommunication; communication of telegraphic or telephonic signals, sound, and/or images by radio frequencies. (2) Empresa Nacional de Telecomunicaciones de Colombia. (3) **Télécom**, satellite operation by France.

teleconferencing   See *video conferencing*.

telefilm   Motion picture film transmitted by television.

telegraph   See *radio telegraph*.

tele-movies   See *made-for-television features*.

Telemundo   Spanish language cable broadcast network (USA).

telenovelas   Popular Spanish language soap operas.

telephone transmission   Special telephone lines carrying electrical impulses at approximately 30,000 miles per second (48,000 kilometers per second) to individual receivers.

telephone channel   See *speech channel*.

telephoto lens   Also called **long lens**. Narrow-angle camera lens with a long focal length used for close-up and distant photography.

teleplay   See (1) *screenplay* or *script*. (2) Dramatic presentation telecast, a television drama.

Teleprompter   Trade name for prompter.

telerecording   British term. See *kinescope recording*.

teletext   One-way text service transmitted by broadcasters that can be decoded by suitable television sets (GB). See also *Videotex*.

Teletype (TTY)   Trade name for a printing telegraph that records like a typewriter. Used in news rooms.

televiewer   One who watches a television show; the television audience.

television (TV)   Electric telecommunication system used for instantaneous transmission of live or recorded sound and visual images in B&W or color, to distant places through the airwaves or cable. It uses VHF and UHF channels with AM for vision and FM for stereo sound.

television black   See *reference black.*

television camera   See *camera/*1.

Television Directorial Awards   Annual awards presentation for direction of television shows and programs in seven categories sponsored by the Directors Guild of America.

television film   Motion picture film produced especially for television viewing.

television home   Survey rating term for households with television sets.

television pick-up tube   Television camera tube that converts the optical images into electrical impulses. See *Image Orthicon, Plumbicon, Vidicon* and *Saticon.*

television receiver   See *receiver.*

television standards   Accepted line/field and color standards for television systems. See *NTSC, PAL, SECAM.*

television star   See *star.*

television studio   See *studio.*

television tube   See *camera tube.*

television white   See *reference white.*

telex   A telegraphic system in which printed signals and messages are sent and received by teleprinters that are connected to public telecommunication network.

Telop   (1) Telopticon; opaque television projector. (2) Opaque photograph or artwork projected by a Telop projector.

Telstar   Domestic communication satellite operated by the American Telephone & Telegraph Company-AT&T (USA).

tempo   The rate of action and speed of a sequence within the overall pace and time of the program or show.

10.5-inch reel   Diameter for a professional magnetic tape reel, equals 28.7cm. See also *7-inch reel* and *14-inch reel.*

10 kHz–100,000 MHz   Range of electromagnetic frequencies. See *radio frequencies.*

tension roller   See *Jockey roller.*

terminal   Fitting for an electrical connection.

terr   Terrestrial.

test   (1) Try-out of a performer for a particular part. (2) Screen test. (3) Make-up or costume test to see how it will appear on the screen. (4) Test film. (5) Voice test.

test bar   See *color bar.*

test card (test pattern)   Linearity chart, composed of geometrical patterns, to aid camera picture alignment. See *gray scale.*

test film   Film designed with standard features for testing picture steadiness, definition, optical alignment, and sound frequency response.

test shot   Camera shot to test a set-up, alignment, or a scene.

test strip   See *test film.*

test tape   An audio or videotape, supplied by the manufacturer, to aid testing and/or alignment of audio and/or video recording and playback equipment.

texture   The impression of depth, surface patterns, etc., on a plane surface, achieved by the use of paint and decorative materials.

TF   Télévision Française (France).

TGO   ITU country code for Togo.

TH   Tungsten-halogen.

THA   (1) ITU country code for Thailand. (2) Turk Haberler Ajansi (Turkey).

Thaumatrope   A revolving disc with an image painted on each side. When rotated, the images appear to be in motion; a simple demonstration of persistence of vision. See *persistence of vision.*

theme   (1) Subject of the program. (2) Recurrent melody of an advertisement (commercial), show, or film.

thermionic valve (also called **vacuum tube**)   A system of electrodes (a cathode, anode or other) arranged in an evacuated glass or metal envelope for electronic circuit applications.

thermoplastic recording (TPR)   Electron beam recording on plastic film via the television camera or other image (symbol) generators, providing instant image development, very high recording bandwidth, color, and high recording density.

13-weeks Broadcast program series broken down four times to 13 weeks, each lasting one quarter of the year; the last being the recess or repeat quarter. See *quarters*.

30-ips Magnetic tape (mastering quality) speed standard, equals 76.2cm/s. Used in older model equipment.

30m Standard film length; equals 100 feet.

30-weeks See *TV season*.

38.1 cm/s Videotape speed standard; equals 15 ips. Used in the 525-line/60-field system.

35mm Professional motion picture film gauge, 35mm in width, with 16 frames to the foot.

39.7 cm/s Videotape speed standard in the 626-line/50-field system, equals 15.625 ips.

33-1/3 rpm Turntable (phonograph) rotation speed standard for long-playing (LP) microgroove records.

3200K Color temperature for tungsten light. See also *5400K*.

threading Proper placement of tape or film in its required path in a recorder, camera, projector, or printing machine.

three-color process Color film reproduction system in which the visible spectrum is divided into red, green, and blue in recording and presentation.

3-D film Three-dimensional film. See *stereoscopic cinema/television*.

3-3/4 ips Recording tape speed standard, equals 9.52 cm/s.

3/8 inch Approximately 1 cm.

three-fold flat A three-part, hinged flat. See also *book/1*.

three-gun tube See *shadow mask tube*.

three-inch lens 75mm lens.

3.58 MHz NTSC chroma. See *chroma*.

3/4 inch Videotape width standard, equals appr. 2 cm (19.05mm), generally known as the **U-matic format.**

3/4 inch with time base corrector Broadcast quality videotape.

366m Film length standard, equals 1,200 feet.

throw See *projection distance*.

through-the-lens See *reflex camera*.

tie-in box Electrical connector box.

TIG Societé des Télécommunications Internationales Gabonaises (Gabon).

tight (1) A program running too close to its allocated time limit. (2) Tight shot; a closely framed performer or subject; extreme close-up (ECU).

tilt The up-and-down movement of the camera along its vertical axis. See also *panoramic shot*.

time Time slot; broadcast time for a program or commercial.

time base corrector (TBC) Electronic device that ensures stable, high quality pictures in videotape recording and reproduction. A necessary device for broadcast or mix/effects use of video signals provided from VHS, Betamax, U-matic and other types of videotape recorders.

time check The synchronization of all time clocks and watches for program timing.

time code Also called **address track.** A data signal recorded on the videotape indicating precise recording of frames, seconds, minutes, and hours specific to a particular address on the tape; SMPTE standards for the United States and Canada, EBU for Europe.

time-lapse cinematography Cinematographic technique whereby single frames (one frame at a time) are shot continuously at a given rate over a period of time to record slow movements, i.e. plant growth. See also *single-frame shooting*.

time out See *break*/1.

time slot Scheduled broadcast period.

time zone (TZ) Standard time of the 24 longitudinal division of the earth's surface, starting from the Greenwich meridian in Great Britain.

timing (1) Determining the exact running time of a show or program. See also *back timing*. (2) Timed print; the grading of the film in the laboratory process. See also *one light print*.

tint See *hue*.

Tiny Mac A compact, accurately focusing Fresnel light that can be concealed or mounted on a stand and atop a camera.

TIO The former Television Information Office of the National Association of Broadcasters (USA). Now defunct.

TIS Traveler's Information (broadcast) Service.

TIT Societé des Télécommunications Internationales du Tchad (Chad).

title Graphic material, lettering, and artwork carrying the main title, production, and performing credits, etc. at the beginning and/or end of a show.

title card See *caption*/1.

title drum See *crawl*.

title music Background music behind the opening and closing titles.

title stand See *animation stand*.

TJK ITU country code for Tadzhikistan.

TKM ITU country code for Turkmenistan.

TLC The Learning Channel cable television network (USA).

TMC The Movie Channel cable television network (USA).

TNN The Nashville Network (USA).

TNT Turner Network Television (USA).

T-number Transmission number. See *T-stop*.

"To black" See *black*/5.

Todd-AO Wide-screen film presentation technique using 65mm camera original and 70mm prints with stereophonic sound on a magnetic track (USA).

toe mark Studio (set) floor mark used to indicate exact place of action and/or equipment. See *mark*/1.

Tokyo International Film Festival Competitive and non-competitive festival for feature films in 35mm and 70mm formats, combined with a film market and conference, held in Tokyo (Japan).

TON ITU country code for Tonga.

tone (1) The strength, pitch and quality of sound. (2) The color of any photographic image; any shade of gray. (3) Test tone recorded on tape for level setting and alignment.

tone control Bass and treble controls in an audio. Amplifier to alter low- and high-frequency characteristics and to compensate for room acoustics and to reduce hiss, hum, and scratching noises.

tongue The horizontal move of a camera mounted on a boom dolly, either from left to right or from right to left.
Called **jibbing** in Great Britain.

toning The altering of the monochrome (B&W) photographic image to another color using a chemical solution, i.e. sepia, to achieve an old (antique), or other effect.

top hat See *high hat*.

top light Light source directly over the subject.

top shot See *high-angle shot*.

top sun The noontime sunlight, the brightest and strongest exterior light.

Toronto International Film Festival/Festival of Festivals Annual film festival held in Toronto, Canada, with an emphasis on new films.

torque motor A motor driving the camera magazine with increasing torque in relation to film load.

TP (1) Telephone. (2) Teleprensa (Colombia).

TPR Thermoplastic (electron beam) recording.

T/R (TV/R) Television/radio.

track (1) See *soundtrack*. (2) Track for dollying shots.

tracking shot See *dolly shot*.

track laying (1) Sound editing process; the editing and synchronization of various soundtracks for a videotape or film. (2) The laying of tracks, metal rails for moving dolly shots.

traffic department Traffic; a broadcast station department responsible for the minute-by-minute daily programming log and commercial placement, and the scheduling of the weekly or daily announcing and technical personnel.

trailer A short promotion film, usually to publicize a theatrical feature.

transcription A recording. See also *electronic transcription*.

transducer A device actuated by power from one system while supplying power to another system in another form. See also *head/2*.

transfer (1) Transfer of copyright; written permission from the copyright owner of a creative work to assign (transfer) the copyright. (2) Sound duplication by recording from the original 1/4 inch magnetic tape to 16mm or 35mm magnetic film; and the final transfer from these larger gauges onto a magnetic or optical duplicating copy to comprise the sound portion of a film.

transfer loss The loss in sound quality during the transfer process—occurs mostly in optical transfer.

transformer   Electrical apparatus for reducing or increasing the voltage of alternating current without changing its frequency.

transistor   A compact size semiconductor (non-vacuum amplifier) to control electric flow, used similarly to the thermionic valve or vacuum tube.

transition   Sound and optical effects used in moving from one scene to another by fades, musical bridges, etc.; carried out in editing.

translator   A self-contained receiver/transmitter placed on higher elevation to extend, to increase the range of television broadcast coverage. See also *transmitter*.

translucent screen   Rear projection screen of transparent plastic or semi-opaque material used in small-scale projection.

transmission   (1) Communication technique for passing on (transmitting) information by powerful radio frequencies (carrier wave) generated in the transmitter, and audio signals modulated (superimposed) upon the carrier wave. (2) The amount of light passing through an object.

transmission controller   Coordinator in the master control responsible for the smooth presentation flow from all broadcast sources.

transmission stop   See *T-stop*.

transmitter   Broadcast equipment for electromagnetic radiation of radio frequencies, comprised of devices to produce the carrier wave, modulate it, and feed it to aerials.

transparency   See *slide*.

transparent tape   See *splicing tape*.

transponders   Units that receive and transmit programs from satellites then amplify and transmit them to earth.

transverse scanning   High-quality, professional videotape recording in the 2-inch format whereby four heads rotate across the tape at high speed, recording the video signal (audio track, video track, control track, cue track) on the tape. See *quadruplex head* and *quadruplex recording*. See also *helical scan*.

transverse waves   Electromagnetic radiation in which the displacement of the waves takes place in a plane surface at right angles to the direction of propagation of the waves.

trapeze   An overhead fixture to attach, hang, and move lights (luminaires) with rope or chain.

traveler curtain   A large curtain that opens in the middle or on the side.

traveling matte Process whereby the foreground action is photographed in front of a large blue screen in the studio and the background is recorded separately. The composite is done by use of filters in an optical printer. In the finished film the actors appear in distant places far away from the studio. See also *back projection/2*.

traveling shot See *dolly shot*.

travelogue A videotape program or a documentary film describing travels, countries, and scenic places, usually combined with narration, commentary, and often canned music.

TRC ITU country code for Tristan da Cunha.

TRD ITU country code for Trinidad and Tobago.

treatment Intermediate step between synopsis and script, with detailed sequences and locations, describing the characters and their interrelated actions and giving the dramatic high points of the dialogue. See also *script* and *synopsis*.

triacetate Cellulose triacetate film base with low flammability (slow burning).

trial print The first composite sample print with pictures and corresponding sound. See *answer print*.

triangle Also called **spreader.** A star-shaped triangular aluminum, plastic, or wooden device that holds the legs of a tripod. It can be folded for compact packing. In Great Britain it is called a **crowfoot.**

trim (1) Adjustment (frame adjustment) in tape editing to either add or subtract time. (2) Pieces of film shots left after the editing assembly of the workprint or finished film. See also *out takes*.

trim bin Also called **editing bin.** A fiber film container with hanging pins, used in the editing room.

Triniscope Large-screen color receiver display tube used in earlier telerecordings. See *Eidophor*.

Trinitron A high-quality television picture tube using only one electron gun and one aperture grid.

tri-pack Color motion picture film type with three layers of sensitive emulsion coating.

tripod Adjustable, three-legged camera support consisting of tripod legs and a movable head.

tripod dolly A portable, three-wheeled triangle dolly.

tripod head   Camera mount. See *head/5, friction head,* or *gyro head,* etc.

tripod scooter   British term for **tripod dolly.**

trip switch   See *buckle switch.*

trolley battery   Wheeled battery pack on a dolly or hand truck. See *battery.*

tromboning   Colloquial for zoom lens technique in small format cinematography with variable speed in movement, acceleration, and deceleration.

trouper   (1) Colloquial term for a follow spotlight utilizing either arc, incandescent, or quartz lamps. (2) Slang for an actor in a road show.

TRT   Turkish Radio Television.

trucking   Also called **crabbing.** Trucking shot; movement of the camera on a dolly or other vehicular support, parallel with the performers, to pace and maintain image size. See *dolly shot.*

TRVO   Television receive-only dish antenna.

TS   Time slot.

TSA   Total service area.

TSCJ   Telecommunications Satellite Corporation of Japan.

TSR   Télévision Suisse Romanche (Switzerland).

T-stop   Also called **true f-stop.** Transmission number or T-number; a lens calibration system that takes the true transmittance of the lens into account free from all absorption loss and reflection. See also *f-stop.*

TSTT   Telecommunications Services of Trinidad & Tobago, Ltd.

TT   Tidningarnas Telegrambyra (Sweden).

TTL   Through-the-lens. See also *BTL* and *reflex camera.*

TTNA   Ta Tao News Agency (Republic of China).

TTY   Teletype.

TUN   ITU country code for Tunisia.

tuned speaker   A specially engineered speaker that produces greater low frequency (bass) output.

tuner   A reception device and tune-in, used to set and find the right wavelength to receive a particular audio, video, or data signal.

tungsten (1) Tungsten-halogen light; incandescent studio light with (Wolfram) tungsten filament, usually balanced for 3000, 3200 or 3400 K. (2) Term for interior light ISO (DIN) rating. See also *daylight*. (3) Tungsten film; color film balanced for tungsten (artificial) light. See *type A film* and *type B film*.

tuning fork A metal fork with two prongs that gives a pure tone of a specified pitch when struck, used in acoustics and for tuning musical instruments.

TUR ITU country code for Turkey.

turntable (1) A circular rotating platform used in the studio. (2) A phonograph record player in the control room.

turret See *lens turret*.

TUV ITU country code for Tuvalu.

TV Television.

TVB (1) Television Bureau of Advertising (USA). (2) Television Broadcast, Ltd. (Hong Kong).

TV Marti Television service to Cuba broadcast by the U.S. Information Agency.

TV mask See *essential area*.

TVR Television (kinescope) recording; kine.

TV/R (T/R) Television/radio.

TV reticle A precision-etched, flat, ground glass used in specific lenses with an outlined television frame (mask) that also gives a full field of view.

TVRI Televisi Republic Indonesia.

TVRO Television receive-only unit.

TV season (television season) A 30-week run of a particular series, usually from mid-September to mid-April. However, new shows are introduced during other times as well.

TVZ Television Zanzibar (Tanzania).

TWC The Weather Channel cable television network (USA).

tweeter A loudspeaker that produces higher frequency (10kHz–17kHz) sounds. See also *woofer*.

1200 ft Film length standard, equals 366m.

TWN ITU country code for Taiwan.

two-color process Obsolete color reproduction system whereby the visible spectrum is divided into blue-green and orange-red. See *three-color process*.

12.7mm Magnetic recording tape width standard, equals 1/2 inch.

25 fps Sound film speed in the European-Continental 625-line/50-field television picture standard.

25 ft 8mm film length standard, equals 7.62m.

25.4mm Magnetic recording tape width standard, equals 1 inch.

24 fps Sound film speed for the North American standard of 525-line/60-field television films. It is the standard projection rate for sound films. See *sound speed*.

24.3 cm/s Magnetic tape speed (professional 1-inch type C videotape), equals 9.6 ips.

22.2°C 72° F temperature; ideal condition for tape and film storage. See *vault*.

two-element tube Diode; Solid state device that allows current flow in one direction only.

two-field picture A complete, interlaced, television picture frame.

two-fold flat See *book/1*.

2 in Magnetic recording tape width standard, equals 50.8mm. Now obsolete.

two-inch lens 50mm lens.

two-shot Photographic shot including two subjects or two performers.

2.35:1 Also called **two three five.** Aspect ratio for a wide-screen cinema process photographed by an anamorphic lens. See also *1.85:1, Cinemascope,* and *Panavision*.

TWR Trans World Radio (USA).

TX Transmitter.

type A film Camera film balanced for 3400 K tungsten light.

type B film Camera film balanced for 3200 K tungsten light.

type C film Broadcast quality format 1 inch videotape.

type G film Camera film with reduced color sensitivity, balanced for mixed light conditions.

TZ Time zone.

TZA ITU country code for Tanzania.

UAE ITU country code for the United Arab Emirates.

UCC Universal Copyright Convention. See *copyright*.

UER Union Européenne de Radiodiffusion; European Broadcasting Union-EBU.

UF Ultrasonic frequency.

UFA University Film Producers Association (USA).

UGA ITU country code for Uganda.

UHF Ultra high frequency.

UHS (UHSS) Ultra high-speed system.

UIT Union Internationale des Télécommunications. See *International Telecommunication Union-ITU*.

UK News United Kingdom News.

UKR ITU country code for the Ukraine.

UKW Ultrakurzwelle; FM band (Germany).

ultra high frequency (UHF) Microwaves of 300–3,000 MHz, used in point-to-point transmission of television signals on channels 14 and above.

Ultra Panavision See *Panavision/3*.

ultrasonic cleaning Effective means of cleaning release prints in a liquid solution agitated by ultrasonic frequencies.

ultrasonic frequency (UF) Frequency in excess of 20,000 Hz.

ultrasonic generator Apparatus used to produce ultrasonic frequency.

ultraviolet filter (UV filter) Camera lens filter designed to select (control) ultraviolet waves present in sunlight and in arc and mercury vapor lamps.

ultraviolet photography   A filming technique using ultraviolet radiation. There are principally two methods employed:
1. Reflected light ultraviolet photography;
2. Back-light (fluorescence) photography.

In motion picture filming, the back-light (fluorescence) method is used and the light source is equipped with special filters, called **exciter filters.**

Both B&W or color film may be used, however, B&W film will produce a better, more dramatic effect.

ultraviolet radiation   Electromagnetic radiation at the very short wavelength of the spectrum, between the visible light waves and X-rays (sun rays), affecting photographic film. A portion of the undesirable ultraviolet radiation can be controlled by filters. See *ultraviolet filter.*

U-matic   Trade name for a Sony-developed ¾ inch format video cassette recorder used in both consumer grade and broadcast quality applications.

umbra   Complete shadow.

umbrella   Studio and location umbrella used for bouncing light and for shading cameras or other equipment.

U.N.   United Nations.

UNDA   International Catholic Association of Radio, Television and Audiovisuals—Association Catolique Internationale pour la Radio, Télévision et l'Audiovisuel (Belgium). Unda=Latin for wave.

"Under"   Also called **"Hold under"** and **"Hold to BG."** Cue to hold effects and music at background levels. See also *background*/1.

undercranking   See *fast motion.*

underexposure   Insufficient exposure due to either inadequate lighting or wrong lens setting. In film it may be a wrong f-stop, incorrect camera speed, or a combination of both. The result may be dark tones, a weak reproduction of the negative, or a dark positive print.

understudy   Replacement actor/actress for a (usually major) part in a teleplay or film. See also *quick study.*

underwater camera   See *underwater housing.*

underwater filming   Cinematography done underwater with a specially equipped camera by a diver-camera operator. Shooting may be accomplished in natural undersea situations or in a studio water tank under controlled conditions.

underwater housing   Underwater camera housing; pressure resistant or pressurized, watertight camera housing with outside controls, used to hold camera in underwater cinematography.

UNESCO   United Nations Educational, Scientific and Cultural Organization.

UNI   United News of India.

unidirectional current   See *direct current.*

unidirectional mike   A directional microphone with a pick-up pattern in one pointed direction.

Union Internacional des Télécommunicacions (UIT)   See *International Telecommunication Union-ITU.*

UNISAT   British Satellite Corporation (operating since 1982).

unit   The crew; the team of technicians and craftsmen/women working together during a production.

United Nations Educational, Scientific and Cultural Organization (UNESCO)   A specialized agency of the United Nations founded in 1946 with the purpose of furthering peace and understanding among the nations through education and a broad range of scientific and cultural programs.

   Information exchange and communication are a crucial part of the activities involving over 170 member countries.

   Headquartered in Paris, France, UNESCO operated offices and divisions include:

   OPI   Office of Public Information;

   AVP   Audio-Visual Production Division;

   IIP   International Informatics Programme;

   COM   Communication Division; and

   IPDC   International Programme for the Development of Communication.

   Publications include the *UNESCO Courier* (in 36 languages and Braille); *UNESCO Sources* (in French, English, Portuguese, and Spanish); several specialist journals, *Statistical Yearbook, World Communication Report,* and series in education, social sciences, and world history.

United States Information Agency (USIA)   An independent agency of the U.S. Executive Branch, it was founded in 1953 and merged with the Bureau of Educational and Cultural Affairs of the Department of State in 1977. Its main function is the dissemination abroad of information about the United States and the organization of cultural and educational exchanges via its offices in 125 countries. USIA maintains the Voice of America, Radio and TV Marti and Radio Free Asia. See also *VOA, Radio Marti, TV Marti* and *RFA*.

United States Office of War Information (OWI)   The Office was created in 1942, during World War II, to consolidate the various government information activities, both domestic and foreign. By 1945 the Office's activities were mostly directed abroad and later its foreign area activities were taken over by the Department of State. OWI was abolished in 1946. See *United States Information Agency*.

unit production manager (UPM)   The person in charge of coordinating the technical unit(s), budgeting, shooting schedule, and location arrangements during a television or film production.

Univision   Spanish language broadcast cable network (USA).

UNRAP   Union de Radioemisoras de Provincias del Peru.

UNRTA (Union of National Radio and Television Organizations of Africa)   Unité des Radios Télévisions Nationales d'Afrique-URTNA (Senegal).

unwinding   See *winding/unwinding*.

UP   Uutispalvelu (Finland).

UPI   United Press International (USA).

up-link   Transmitter/transmitting dish to orbiting satellite; ground-to-satellite transmission. See also *down-link*.

UPM   Unit production manager.

UPP   United Press of Pakistan.

URG   ITU country code for Uruguay.

URS   ITU country code for the former USSR. Now obsolete.

See individual country codes of the Commonwealth of Independent States.

URTNA (Unité des Radios Télévisions Nationales d'Afrique)   Union of National Radio and Television Organizations of Africa-UNRTA (Senegal).

US News Wire   Wire service distributing press releases of clients for a fee to the media.

USA   (1) ITU country code for the United States. (2) United Scenic Artists (USA). (3) USA Network cable television channel.

USAFRTS   See *AFRTS*.

U.S.A.S.I.   United States of America Standards Institute; name of the former American Standards Association (ASA).

USCI   United Satellite Communications, Inc. (USA).

USIA   United States Information Agency.

USIS   United States Information Service; called USIA abroad.

UTC   Universal Time Coordinate/Coordinated Universal Time.

UV   Ultraviolet.

UV filter   See *ultraviolet filter.*

UV photography   See *ultraviolet photography.*

UZB   ITU country code for Uzbekistan.

V (1) Volt. (2) Vertical (polarization).

VAC Volts alternating current.

vacuum tube Thermionic valve.

valve Radio tube.

van See (1) *Cinema van*. (2) *Remote van*.

variable area recording Method of optical sound recording producing a laterally divided oscillographic trace, a variable area track.

variable density recording A method of optical sound recording whereby the signal forms narrow bands of density graduations in the full width of the track.

variable focal-length lens Also called **varifocal lens**. Zoom lens.

variable shutter A rotating mechanical device between the lens and the film in a camera, with a variable opening that blanks out film in an intermittent motion. See also *shutter*.

variable speed motor See *wild motor*.

VariCon Variable contrast filter; a compact variable contrast control system that provides a continuously adjustable contrast over the entire photometric range. See also *pre-flashing, post-flashing* and *Panaflasher*.

variety program See also *comedy/variety program*.

vault Climatized, fireproof safe storage for magnetic tape and film material. Ideal storage conditions for raw color stock are
55°F (13°C) for up to 6-month period; Storage over 6 months should be at 0°F to 10°F (−18° to −23°C). After removal from cold storage the sealed tape or film must be allowed to warm up to room temperature (appr. 70°F [20°C]), before being opened and used.

VBS Video blanking sync.

VC   Video cartridge, video cassette.

VCR   Video cassette recorder.

VCT   ITU country code for St. Vincent and Grenadines.

VDC   Volts direct current.

VDR   Video disc recorder.

VDT   Visual display terminal.

vectograph   A three-dimensional effect achieved by two superimposed images in a composite picture viewed through polarized lenses. See *stereoscopic picture*.

vectorscope   A special oscilloscope that checks the color accuracy of video cameras, switchers, VCRs and other equipment, displaying information related only to chromaticity.

veejay   Video jockey.

velocity mike   See *ribbon mike*.

velocity of light   The speed of light, 186,282 miles/sec., or 298,051 kilometers/sec. (mean value).

VEN   ITU country code for Venezuela.

Venice International Film Festival   See *Biennale di Venezia*.

VENPRES   Government-controlled and -run news agency of Venezuela.

vertical cut   See *straight cut*.

very high frequency (VHF)   Frequencies between 30–300 MHz broadcast on television channels 2 to 13.

very low frequency (VLF)   Frequencies between 3–30 kHz with a wavelength of 10 km.

VESA   Video Electronics Standard Association (International).

VF   (1) Varying frequency. (2) Viewfinder.

VHD   (1)Video high density. (2)Video disc system.

VHF   Very high frequency.

VH1   Video Hits One cable television network (USA).

VHS   Video home system.

video   (1) The picture part of television transmission. (2) Short videotape, referring mostly to musical (rock) production or promotion.

video assist   Attachment to a motion picture film camera to record video images simultaneously; used to facilitate (assist) framing, picture composition, also as an editing aid. See also *System 35.*

video black   See *black/3.*

video blanking sync (VBS)   A composite video sync, either NTSC, PAL or SECAM, that incorporates all components needed for viewing, recording, and transmission.

"Video blues"   Stress related condition of heavy VDT users.

VIDEOBRASIL   International Video Festival in Arts, Music, Entertainment, Documentary and general categories, held annually in São Paulo (Brazil).

video cable   See *coaxial cable.*

video camera   See *camera/1.*

video cartridge   A plastic videotape container housing a single reel for special recording and playback with short program segments. No longer in use.

video cassette (VC)   A plastic cassette, similar to sound recording cassette, in which a supply and take-up reel is housed for recording and playback of television shorts, programs, and films transferred onto videotape.

video cassette recorder (VCR)   A recorder that accepts a cassette in which the videotape is housed. A variety of video cameras contain built-in VCRs. See also *camcorder* and *videotape recorder.*

video cassette recording   Video recording onto a tape housed in a cassette (as opposed to open reel recording). See also *videotape recording.*

video conferencing   Formerly called **teleconferencing.** Conference meeting procedure from two or more (distant) locations using low-cost telephone and data cables or satellite-transmitted full-motion color television, now using special encoder/decoders called **CODECs.**

video control panel   See *console/1.*

video disc   A disc, similar to an audio disc, that records and plays back video and sound signals. It can hold more information than tape, with easier search functions and better endurance than tape. See also *disc* and *compact disc.*

video disc recorder (VDR)   An electronic device that records and plays back video and sound signals on video discs, with instant playback, freeze-frame, and slow-motion capabilities.

video editing　See *videotape editing*.

video 8　See *8mm/*1.

video engineer　Technical personnel in charge of electronic maintenance of the television equipment. He/she adjusts the exposure and total quality of the picture and matches each picture produced by the cameras, telecine, and other image sources.

videographer　Colloquial for *video camera operator*.

video home system (VHS)　1/2-inch video cassette format for consumer (home) use.

video jockey　Veejay; television disc jockey.

video monitor　See *monitor/*2. See also *video assist*.

video news release (VNR)　News-type advertising of products and consumer goods released as "news."

video noise　See *noise/*2. See also *snow*.

video recording system　A device using magnetic tape or discs capable of recording and reproducing video and audio signals.

video signal　Electrical impulses conveying the visual information in television. See also *signal/*1.

video sunglasses (Glasses cam)　A micro video camera concealed in sunglasses and used in undercover investigative reporting, and increasingly by intelligence agencies. The pair of ordinary-looking sunglasses contains a built-in micro-lens in the nose bridge (giving an exact eye-level view), an image sensor (CCD), a tiny amplified microphone, and a NiCad battery. A small cable runs in the back (under a shirt or blouse) to a microrecorder the size of a small paperback book. See also *point-of-view camera*.

video switcher　See *vision mixer/*2.

videotape (VT)　See *magnetic tape*.

videotape editing　(1) Physical editing; cutting and splicing the tape as in film editing. Now obsolete. (2) See *electronic editing*.

videotape leader　Leader with audio and video information attached to the beginning of the videotape program, carrying 10 sec. color bars, 15 sec slate information, 8 sec. Academy numbers, and 2 sec. black (SMPTE).

videotape recorder (VTR)　Video recorder with open reel tape. The first videotape recorder and tape was developed at the AMPEX Corporation by Charles I. Ginsburg (1920–91) and his six-man team in 1956. See also *video cassette recorder*.

videotape recording (VTR)   Electronic recording of a performance, show, or television program whereby scenes taken by the television cameras are recorded on a separate videotape recorder and may be stored, duplicated, packaged, or played back instantaneously or at a later time.

Videotex   Colloquial for a screen-based interactive information system. See also *teletext* and *viewdata*.

video track   The television picture recorded on a videotape.

Vidicon   A durable and compact photo conductive camera tube used for consumer grade cameras and in telecine, industrial, educational, and closed-circuit operations. See also *Image Orthicon, Plumbicon,* and *Saticon.*

VIENNALE   Vienna International Film Festival; Non-competitive yearly feature film festival held in Vienna (Austria).

viewdata   A form of Videotex transmitting information from a central computer, offering a two-way interaction service via the television screen for pictures and the public telephone network for sound. See also *CEEFAX.*

viewer   (1) Device used in the film editing room on which film can be viewed in a magnifying ground glass. The film is run through the viewer with the aid of rewinds. (2) Visual inspector, operator in a film processing laboratory. (3) Person watching a television or film program or show; member of the viewing audience.

viewfinder   (1) Flat-faced small picture monitor above the television camera tube enabling the camera operator to compose, frame, and focus an image. (2) An optical device mounted on (as a direct viewfinder) or incorporated in the film camera (TTL) for viewing a scene to be photographed.

viewing filter   See *color contrast viewing filter* and *panchromatic viewing filter.*

vignette   A picture with a sharp, detailed center but blurred edges that may be the result of a faulty lens, or introduced as a visual effect.

VI meter   Volume indicator; a meter with volume unit reading. See *VU meter.*

VIR   ITU country code for the U.S. Virgin Islands.

VIRS   Vertical interval reference system.

virtual focus   See *focal point.*

virtual image photography   See *aerial image cinematography.*

vision mixer　(1) Video control panel. (2) Video switcher. British term for the **video control operator.**

Vistavision　(1) Wide-screen film production system that used an optically reduced print from a double-sized frame 35mm negative. (2) Vista Vision; trade name for 35mm film cameras.

visual　The image portion of the television or film program.

visual aids　Maps, charts, graphs, slides, pictures, films, and three-dimensional models used to enhance a visual presentation.

visual effects　See *optical effects.*

visual image cinematography　See *aerial image cinematography.*

visualizer　British term for **cue card.**

visual medium　Television, film, and/or photography.

visuals　Still photographs and slides used as part of a television or film project.

VIZ　Abbreviation for *visual (video).*

VJ　Video jockey. Also called **veejay.** See also *DJ.*

VLF　Very low frequency.

VLS　Very long shot.

VNA　Vietnam News Agency.

VNF　Video news film; fast film, sensitive to low light.

VNR　Video news release.

VO　Voice-over; off-screen narration or commentary. See *narration.*

VOA　Voice of America; broadcast service of the U.S. Information Agency transmitted in several languages and only to audiences abroad (outside the U.S.).

VOD　Video-on-demand.

voice test　The testing of a performer's voice quality over the microphone and his/her ability to speak, act or sing.

voice slating　Sound slating, in lieu of a clapper board, by either hand clapping, quick sound, or tapping at the microphone.

volt (V)　The international system unit of electric potential and electrical force that is capable of carrying one ampere of current against the resistance of one ohm. Named after Conte Alessandro Volta (1745–1827), Italian physicist and professor.

voltage   The supply of electricity measured in volts.

voltage regulator   Device that compensates for fluctuation in an electrical current.

volt meter   An instrument for measuring the potential difference between two points of an electrical circuit in volts.

volume   Sound intensity, amplitude within the audio range.

volume control   See *fader.*

VOR   Voice-over recording.

VR   Vatican Radio.

VRG   ITU country code for the British Virgin Islands.

VT   Videotape.

VTN   ITU country code for Vietnam.

VTR   Videotape recorder; videotape recording

VTR unit   Videotape recording equipment; a compact unit with a camera, monitor, microphone(s), recorder, videotape, batteries, and lights.

VU   Volume unit.

Vue sur les Docs   See *European Documentary Film Festival.*

VU-graph   Overhead transparency projector. Obsolete term.

VU meter   Metering device indicating the volume of transmitted sound by volume units (VU) in decibels (dB) from −20 to +3.

VUT   ITU country code for Vanuatu.

VZ   La Voix du Zaire.

W (1) Watt. (2) The first letter of the call letters of broadcast stations east of the Mississippi, with the exception of KDKA in Pittsburgh, and KYW in Philadelphia, Pennsylvania (USA).

WACC World Association of Christian Communication (GB).

WAL ITU country code for Wallis and Futuna Islands.

walkie talkie Also called **walkie-T.** Colloquial term for a pair of portable, battery-operated transmitting-receiving sets used in wireless (talk) communication.

walk-on A non-speaking role. See *extra*.

walk-through A television or film rehearsal where production and technical personnel perform, pace, and "walk through" the principal actions. It usually precedes the camera rehearsal. Called **stagger-through** in Great Britain. See also *dry run*.

wardrobe See *costume*.

warm-up Introductions and preparation (getting into mood) of a live audience in the broadcast studio prior to transmission.

warm-up time See *line-up time*.

wash Washing; the bath in a film processing machine that removes, washes off the remaining chemicals from the film after developing and fixing.

water spots Water marks, spots, left on the finished film surface after processing and drying.

watt (W) Unit of electrical power. A steady electrical current flowing through the ends of a conductor at a potential difference of 1 volt. Named after James Watt (1736–1819), Scottish inventor.

wattage Electrical power measured in watts.

watt meter Instrument for the direct measurement in watts of electrical power.

wave   See *electromagnetic waves.*

waveform   The shape of wave motion.

waveform monitor (WFM)   Cathode ray oscilloscope that indicates the video signals and their characteristics.

waveform oscilloscope(WFO)   Cathode ray oscilloscope used to measure waveforms.

wave guide   A conducting tube for microwave signals.

wavelength   The distance between two successive wave points of equal phase in the line of advance. Wavelength is measured by the velocity of the wave motion divided by its frequency.

wave meter   An instrument for measuring wavelength.

wave motion   The propagation of the wave forward at a distance equal to its wavelength.

wax pencil   See *marking pencil.*

WBU   World Broadcasting Unions. See *ABU, AIR, ASBU, EBU, NANBA, OTI, URTNA.*

WDR   Westdeutscher Rundfunk (Germany). See *ARD.*

weave   Undesirable lateral motion of the film in a camera or projector.

wedge   An angled metal device fixed to a tripod head or camera mount to provide a specific shot angle.

Weston meter   Photoelectric exposure meter for measuring Weston emulsion speed. Now obsolete. Named after Edward Weston (1850–1936), American electrical engineer.

wet carrel   Colloquial term for a study carrel equipped with audiovisual hardware. See also *dry carrel.*

wet gate printer   A specific printer in the processing laboratory where the film is immersed in liquid to minimize the effect of fine scratches.

WFM   Waveform monitor.

WGA   Writers Guild of America.

WGC   Writers Guild of Canada.

WGGB   Writers Guild of Great Britain.

WHA   See *experimental radio.*

whip pan   See *swish pan.*

white bar Lines of peak white used as a test signal.

white level White balance; the brightest part, full (peak) white television signal level. See also *black level.*

white reference See *reference white.*

wide-angle lens Short focal-length lens covering a large field of view. See also *long focal-length lens.*

wide-angle shot Camera shot, usually a long shot (LS), taken with a wide-angle lens, covering a large angle of view.

wide band Wide-band cable; capable of carrying signals from several stations.

widescope An old-style wide-screen presentation technique that used two projectors throwing images onto two side-by-side screens.

wide-screen process Film presentation technique of various types in a ratio of 1.65:1 and 1.85:1, aimed to cover a larger view than the normal screen aspect ratio of 4:5. (The normal ratio of 4:3 represents only about 30% of what the human eye is capable of seeing). See also *Cinemascope* and *Cinerama.*

wild motor A variable speed film camera motor that is not subject to precise speed (fps) control, used mainly in silent (non-sync), i.e. "wild" filming.

wild recording Wild track; Nonsynchronous recording of mostly sound effects and background sounds made during the filming. See also *room tone.*

winding/unwinding The rolling and transfer of a tape or film from one reel or core to another with continuous rewinds.

winding A & B See *A wind* and *B wind.*

wind machine A large, specially constructed fan to create wind or wind effects.

window dub Workprint equivalent of videotape for editing purposes.

windscreen A lightweight, soft, microphone shield for outdoor use.

wind-up Cue to the talent to conclude, to indicate that the program is coming to an end.

winging Directing a program without previous rehearsals.

Winton tripod An old-style tripod used by the renowned documentary filmmaker Robert Flaherty.

wipe (1) Optical effect in a variety of shapes that serves as transition between two successive shots, achieved in television by electronic means, and in film with a laboratory process. Brazilian director Alberto Cavalcanti was the first to use wipe in a non-theatrical film in his "Rien que les Heures" (made in France) in 1926. (2) Wipe, erase a recording on magnetic tape.

wireless   Old term for radio, also used for telegraph or telegram. See *radio* and *radio telegraph*.

wireless mike   Also called **radio microphone**. A radio microphone system with a compact, powerful transmitter carried and concealed by an actor or speaker that transmits to a wireless receiver some distance away.

wire service   News service.

wire slug   See *slug*/1.

wobble effect   A heat haze effect achieved by holding a heat distorted plastic sheet in front of the camera lens.

womp   See *hot spot*.

woofer   A loudspeaker designed to reproduce lower audio frequency sounds. See also *subwoofer* and *tweeter*.

wooliness   Excessive reverberation.

work for hire   Work assigned to staff or free-lance writers and composers for payment or an agreed upon fee. In these cases the employer is considered to be the "author," therefore the copyright owner.

workprint   (1) A videotape reference copy. (2) A cutting copy, a positive copy of the camera film original, used as the editor's working film reel. The workprint is assembled (rough cut, fine cut) to guide negative cutting. See also *fine cut, rough cut, negative assembly*.

World Association of Community Radio Broadcasters, The   Association Mondial de la Radio Communauté-AMARC; a non-governmental organization serving community radios worldwide, with a Secretariat in Montreal, Canada, and regional offices in Europe (Sheffield, England) and Latin America (Lima, Peru). AMARC publishes the *InteRadio* newsletter.

World Festival of Animated Films   Annual animated film festival held in Zagreb, Croatia, honoring six categories.

World Film Festival—Montreal   See *Festival des Films du Monde—Montréal*.

Worldnet   A daily television service run by the U.S. Information Agency via satellite to foreign broadcasters.

wow   Slow sound frequency deviation in recording and reproduction with reels of magnetic tape. See also *flutter.*

WR   Weather report.

wrap   "It's a wrap." Wrap up; command to finish the production, strike (or rake) the set, and lock or pack equipment.

wratten filter   An optically corrected gelatin (gel) filter, used in addition to or in place of a glass camera lens filter.

WRBB   World Radio Bible Broadcasts (USA).

writer   See *screenwriter.*

Writers Guild of America (WGA)   A union of radio, screen, and television writers operating in two divisions on both sides of the Mississippi: Writers Guild of America-East (New York) and Writers Guild of America-West (Los Angeles).In the early days the radio writers (of the 1930's) and the screenwriters (of 1933) were members of the Authors League. In 1954, at the dawn of television, the radio and screen writers withdrew from the Authors League and founded the Writers Guild of America-East and -West, while the Authors League retained the Dramatists Guild and the Authors Guild.

The members of WGA are free-lance and staff writers in various categories. Staff writers include news writers, news editors, desk assistants, graphic artists, continuity writers, on-air promotion writers, researchers, news production assistants, production assistants, news assignment desk staff, viewers, and producers. Free-lance writers include motion picture screenwriters, TV dramatic writers, TV episodic series writers, TV daytime serial writers, TV comedy-variety writers, children's TV writers, radio and TV documentary and public affairs writers, and radio comedy writers.

The East and West elect their own officers, Council and Board respectively, and a National Council coordinates the activities of both divisions.

writing beam   Electron beam in a cathode ray oscilloscope producing the image. See also *pilot beam.*

WTN   Worldwide Television News (GB).

WWJ   See *experimental radio/2.*

XCU Extreme close up—ECU.

xenon lamp Xenon arc lamp; a long-life, high-output projector lamp used in place of a carbon arc lamp. See also *arc/*1.

XINHUA Xinhua News Agency (People's Republic of China).

XL Existing light.

XLR Type of small connectors used for audio connections on microphones and portable equipment: 3-pin XLR connectors are audio connectors, standardized worldwide; 4-pin XLR connectors are worldwide standards for 12VDC power connections for portable audio/video equipment; 5-,6-, and 7-pin XLR connectors are for special applications.

y   Yellow.

yard   British linear measure, 3 feet or 36 inches; equals 0.9144m.

yellow   Blue-absorbing (minus blue) subtractive primary color.

YEM   ITU country code for Yemen.

YUG   ITU country code for Yugoslavia.

Yuri   Satellite (series) operated by the National Space Development Agency—NASDA since 1978 (Japan).

Yutel   Yugoslav television.

Z

Z   Zone.

ZAI   ITU country code for Zaire.

ZANA   Zambia News Agency.

ZAP   Zachodnia Agencja Prasowa (Poland).

ZBC   Zimbabwe Broadcasting Corporation.

ZDF   Zweites Deutsche Fernsehen (Germany).

zip pan   See *swish pan*.

ZMB   ITU country code for Zambia.

ZNBC   Zambia National Broadcasting Corporation.

zoom   Also called **zoom in** and **zoom out.** Zooming; the gradual changing of the focal length of a lens in a slow or rapid movement, giving the effect of dollying, although the camera remains stationary.

Zoomar   Brand name for a zoom lens.

zoom effect   Zooming achieved electronically on videotape with digital video effects.

zoom control   Manual or electric control of a zoom lens to effect smooth zoom movement. The zoom control may be direct/mechanical or remote, via miniature electric motors. See also *servo focus system* and *zoom lens motor*.

zoom lens   Special camera lens with a variable focal length.

zoom lens motor   A small electrical motor that operates the zoom control to facilitate zoom movements, zoom in or zoom out.

zoom ratio (zooming ratio)   The ratio of a zoom lens from its narrowest (ECU) to its widest (longest) (ELS) setting. e.g. 6:1 (12.5 to 75mm), 10:1 (9.5 to 95mm, or 12 to 120mm), 20:1 (12 to 240mm), also 40:1, 55:1 and greater.

zoom stand   See *animation stand*.

zoopraxinoscope   Forerunner of today's projector. An early mechanism consisting of a lamp house and lens with a revolving disc that projected photographs in a rapid sequence. It was developed by Jean Louis Meissonier (1815–1891), a French genre painter.

zootrope   A similar device to the phenakistoscope with two cylinders. The spinning images on the inside cylinder could be viewed through slits on the outside cylinder.

ZWE   ITU country code for Zimbabwe.

Appendix A:
English–Metric Conversion Tables

Magnetic Tape Width Standards

$\frac{1}{16}$ inch	4.0 mm	(.4 cm)
$\frac{1}{4}$ inch	6.35 mm	(.63 cm)
$\frac{1}{2}$ inch	12.7 mm	(1.27 cm)
$\frac{3}{4}$ inch	19.05 mm	(1.90 cm)
1 inch	25.4 mm	(2.54 cm)
2 inches	50.8 mm	(5.08 cm)

Magnetic Tape Speed Standards

$\frac{15}{16}$ ips	"Books for the Blind"
$1\frac{7}{8}$ ips	4.76 cm/s
$3\frac{3}{4}$ ips	9.52 cm/s
$7\frac{1}{2}$ ips	19.05 cm/s
9.6 ips	24.3 cm/s
15 ips	38.1 cm/s
30 ips	76.2 cm/s

Phonograph Speeds

$16\frac{2}{3}$ rpm
$33\frac{1}{3}$ rpm
45 rpm
78 rpm

Film Length Standards

100 ft	30 m
200 ft	61 m
400 ft	122 m
1000 ft	305 m in 35 mm
1200 ft	366 m in 16 mm

Negative, Sound, Print and Television Recording Films

1000 ft	305 m
2000 ft	610 m
3000 ft	914 m
4000 ft	1219 m

Film Width Standards

Super 8 mm	has 72 frames/ft
16 mm	has 40 frames/ft
Super 16 mm	has 40 frames/ft
17.5 mm	has 16 frames/ft
35 mm	has 16 frames/ft
65 mm	has 12.8 frames/ft
70 mm	has 12.8 frames/ft

Film Projection Speed

Silent Film	16 or 18 fps
Sound Film	24 fps
European-Continental television picture standard	25 fps

Appendix B:
Table of Frequencies

Frequency Bands

VLF	3–30 kHz
LF	30–300 kHz
MF	300–3,000 kHz
HF	3–30 MHz
VHF	30–300 MHz
UHF	300–3,000 MHz
SHF	3,000–30,000 MHz
EHF	30,000–300,000 MHz

Radio Bands

Long wave	LF	150–285 kHz	1,050–2,000 m
Medium wave	MF (AM)	530–1,600 kHz	187–571 m
Shortwave	HF	3–26 MHz	11–120 m
	SW	5–12 MHz	41–50 m
	SW	12–23 MHz	19–41 m
Ultra shortwave	VHF	65–108 MHz	
	former USSR-EE	65–73 MHz	
	FM	88–108 MHz	
	World wide	88–100 MHz	
	West	100–108 MHz	

S-Band	2,535–2,655 MHz
C-Band	3,700–4,200 MHz (USA and Russia)
Ku1-Band	10.7–11.75 GHz
Ku2-Band	11.75–12.5 GHz
Ku3-Band	12.5–12.75 GHz
Ka-Band	18.00–20.00 GHz

Appendix C: Television
Channels, Standards, and Systems

Television Channels

VHF Channels	2–6	Low Band VHF
	7–13	High Band VHF
UHF Channels	14–70	
	71–83	Reassigned for Land/Mobile use (USA)

Existing Television Picture Standards

North American	525-line/60-field	NTSC
European/Continental	625-line/50-field	PAL, SECAM

Proposed Television Picture Standards: HDTV

1,125-line/60-field	Japan
1,250-line/50-field	Europe
1,050-line/60-field	USA
875-line/60-field	Zenith

Color Television Systems

NTSC	National Television Systems Committee	NA
PAL	Phase Alternating Line	E/C
SECAM	Sequentiel Couleur à Memoire	FR-EE-FA

Appendix D: Film Emulsion Speed Conversion Table

ISO	DIN	BSI, JSA
6	9	6
8	10	8
10	11	10
12	12	12
16	13	16
20	14	20
25	15	25
32	16	32
40	17	40
50	18	50
64	19	64
80	20	80
100	21	100
125	22	125
160	23	160
200	24	200
250	25	250
320	26	320
400	27	400
500	28	500
650	29	650
800	30	800
1000	31	1000
2000	34	2000
3200	36	3200

Appendix E:
Television Systems Worldwide

Country	Line/Field	Color
Afghanistan, Republic of	625/50	PAL/SECAM
Albania, People's Republic of	625/50	PAL
Algeria, Democratic and Popular Republic of	625/50	PAL
Andorra, Principality of	625/50	PAL
Angola, People's Republic of	625/50	PAL
Antigua and Barbuda	525/60	NTSC
Argentine Republic	625/50	PAL
Armenia	625/50	SECAM
Aruba	525/60	NTSC
Australia, Commonwealth of	625/50	PAL
Austria, Republic of	625/50	PAL
Azerbaijan	625/50	SECAM
Azores	525/60	PAL
Bahamas, The	525/60	NTSC
Bahrain, State of	625/50	PAL
Bangladesh, People's Republic of	625/50	PAL
Barbados	525/60	NTSC
Belarus	625/50	SECAM
Belgium, Kingdom of	625/50	PAL
Belize	525/60	NTSC
Benin, People's Republic of	625/50	SECAM
Bermuda	525/60	NTSC
Bhutan, Kingdom of	625/50	PAL
Bolivia, Republic of	625/50	NTSC
Bosnia and Herzegovina	525/60	PAL
Botswana, Republic of	625/50	SECAM
Brazil, Federative Republic of	525/60	PAL
Brunei, Darussalam State of	625/50	PAL
Bulgaria, People's Republic of	625/50	SECAM
Burkina Faso	625/50	SECAM
Burundi, Republic of	625/50	SECAM
Cambodia, Republic of	625/50	PAL
Cameroon, Republic of	625/50	PAL

Country	Line/Field	Color
Canada	525/60	NTSC
Capo Verde, Republic of	625/50	PAL
Central African Republic	625/50	SECAM
Chad, Republic of	625/50	SECAM
Chile, Republic of	525/60	NTSC
China, People's Republic of	625/50	PAL
China, Republic of (Taiwan)	525/60	NTSC
Colombia, Republic of	525/60	NTSC
Congo, People's Republic of	625/50	SECAM
Costa Rica, Republic of	525/60	NTSC
Côte d'Ivoire	625/50	SECAM
Croatia	625/50	PAL
Cuba, Republic of	525/60	NTSC
Cyprus, Republic of	625/50	PAL
Czech Republic	625/50	SECAM
Denmark, Kingdom of	625/50	PAL
Djibouti, Republic of	625/50	SECAM
Dominica, Commonwealth of	625/50	NTSC
Dominican Republic	525/60	NTSC
Ecuador, Republic of	525/60	NTSC
Egypt, Arab Republic of	525/60	PAL
El Salvador, Republic of	525/60	NTSC
Equatorial Guinea, Republic of	625/50	SECAM
Eritrea	625/50	PAL
Estonia, Republic of	625/50	PAL/SECAM
Ethiopia, People's Democratic Republic of	625/50	PAL
Fiji, Republic of	625/50	NTSC
Finland, Republic of	625/50	PAL
French Republic	625/50	SECAM
Gabonese Republic	625/50	SECAM
Gambia, The	625/50	PAL
Georgia	625/50	SECAM
Germany, Federal Republic of	625/50	PAL
Ghana, Republic of	625/50	PAL
Gibraltar	625/50	PAL
Greece, Hellenic Republic of	625/50	SECAM
Grenada, State of	625/50	NTSC
Guam (USA)	525/60	NTSC
Guatemala, Republic of	525/60	NTSC
Guinea, Republic of	625/50	PAL
Guinea-Bissau, Republic of	625/50	PAL
Guyana, Cooperative Republic of	625/50	SECAM
Haiti, Republic of	625/50/60	NTSC

Country	Line/Field	Color
Honduras, Republic of	525/60	NTSC
Hong Kong	625/60	PAL
Hungary, Republic of	625/50	PAL
Iceland, Republic of	625/50	PAL
India, Republic of	625/50	PAL
Indonesia, Republic of	625/50	PAL
Iran, Islamic Republic of	625/50	SECAM
Iraq, Republic of	625/50	SECAM
Ireland	625/50	PAL
Israel, State of	625/50	PAL
Italian Republic	625/50	PAL
Jamaica	525/60	NTSC
Japan	525/60	NTSC
Jordan, Hashemite Kingdom of	625/50	PAL
Kalaallit Nunaat	625/50	PAL
Kazakhstan	625/50	SECAM
Kenya, Republic of	625/50	PAL
Korea, Democratic People's Republic	625/60	PAL/NTSC
Korea, Republic of	525/50	NTSC
Kuwait, State of	625/50	PAL
Kyrgizstan	625/50	SECAM
Laos, Lao People's Democratic Republic	625/50	PAL
Latvia	625/50	SECAM
Lebanon, Republic of	625/50	SECAM
Lesotho, Kingdom of	625/50	PAL
Liberia, Republic of	625/50	PAL
Libya, Socialist People's Libyan Arab Jamahiriya	625/50	PAL
Liechtenstein, Principality of	625/50	PAL
Lithuania	625/50	SECAM
Luxembourg, Grand Duchy of	625/50	PAL/SECAM
Macau	625/50	PAL
Macedonia	625/50	PAL
Madagascar, Democratic Republic of	625/50	SECAM
Madeira	625/50	PAL
Malawi, Republic of	625/50	
Malaysia	625/50	PAL
Maldives, Republic of	625/50	PAL
Mali, Republic of	625/50	SECAM
Malta, Republic of	625/50	PAL
Martinique	625/50	SECAM
Mauritania, Islamic Republic of	625/50	SECAM
Mauritius	625/50	SECAM
Mexico, United Mexican States	525/60	NTSC

Country	Line/Field	Color
Micronesia	525/60	NTSC
Moldova	625/50	SECAM
Monaco, Principality of	625/50	PAL/SECAM
Mongolian People's Republic	625/50	SECAM
Morocco, Kingdom of	625/50	SECAM
Mozambique, People's Republic of	625/50	PAL
Myanmar, Union of	525/60	NTSC
Namibia, Republic of	625/50	PAL
Nauru, Republic of	625/50	PAL
Nepal, Kingdom of	625/50	PAL
Netherlands, Kingdom of the	625/50	PAL
New Zealand	625/50	PAL
Nicaragua, Republic of	525/60	NTSC
Niger, Republic of	625/50	SECAM
Nigeria, Federal Republic of	625/50	PAL
Norway, Kingdom of	625/50	PAL
Oman, Sultanate of	625/50	PAL
Pakistan, Islamic Republic of	625/50	PAL
Panama, Republic of	525/60	NTSC
Papua New Guinea	625/50	PAL
Paraguay, Republic of	625/50	PAL
Peru, Republic of	525/60	NTSC
Philippines, Republic of the	525/60	NTSC
Poland, Republic of	625/50	PAL
Portugal, Republic of	625/50	PAL
Puerto Rico, Commonwealth of	525/60	NTSC
Qatar, State of	625/50	PAL
Romania, Socialist Republic of	625/50	PAL
Russia	625/50	SECAM
Rwanda, Republic of	625/50	PAL
St. Christopher and Nevis,Federation of	525/60	NTSC
St. Lucia	625/50	NTSC
St. Vincent and the Grenadines	525/60	NTSC
Samoa, Eastern	525/60	NTSC
Samoa, Independent State of Western	625/50	PAL
San Marino, Most Serene Republic of	625/50	PAL
São Tome e Principe, Democratic Republic of	625/50	PAL
Saudi Arabia, Kingdom of	625/50	SECAM/PAL
Senegal, Republic of	625/50	SECAM
Seychelles, Republic of	625/50	PAL
Sierra Leone, Republic of	625/50	PAL
Singapore, Republic of	625/50	PAL
Slovakia	625/50	PAL/SECAM

Country	Line/Field	Color
Slovenia	625/50	PAL
Solomon Islands	625/50	
Somalia, Democratic Republic	625/50	PAL
South Africa, Republic of	625/50	PAL
Spain	625/50	PAL
Sri Lanka, Democratic Socialist		
Republic of	625/50	PAL
Sudan, Democratic Republic of	625/50	PAL
Suriname, Republic of	525/60	NTSC
Swaziland, Kingdom of	625/50	PAL
Sweden, Kingdom of	625/50	PAL
Swiss Federation	625/50	PAL
Syrian Arab Republic	625/50	PAL
Tajikistan	625/50	SECAM
Tanzania, United Republic of	625/50	PAL
Thailand, Kingdom of	625/50	PAL
Togo, Republic of	625/50	SECAM
Tonga, Kingdom of	525/60	NTSC
Trinidad and Tobago, Republic of	525/60	NTSC
Tunisia, Republic of	625/50	SECAM
Turkey, Republic of	625/50	PAL
Turkmenistan	625/50	SECAM
Uganda, Republic of	625/50	PAL
Ukraine	625/50	SECAM
United Arab Emirates	625/50	PAL
United Kingdom of Great Britain		
and Northern Ireland	625/50	PAL
United States of America	525/60	NTSC
Uruguay, Oriental Republic of	625/50	PAL
Uzbekistan	625/50	SECAM
Vatican City, The Holy See	625/50	PAL
Venezuela, Republic of	525/60	NTSC
Vietnam, Socialist Republic of	525/60	NTSC/SECAM
Virgin Islands, British	525/60	NTSC
Virgin Islands, U.S.	525/60	NTSC
Yemen, Republic of	625/50	PAL/NTSC
Yugoslavia	625/50	PAL
Zaire, Republic of	625/50	SECAM
Zambia, Republic of	625/50	PAL
Zimbabwe	625/50	PAL

Appendix F: National and International News Agencies

AA Agence d'Athenes, Athens, Greece, Hellenic Republic

AA Agencias Andinas, Lima, Republic of Peru

AANS Argus African News Service, Salisbury, Zimbabwe

AAP Australian Associated Press, Sydney, Commonwealth of Australia

A & B Press Monrovia, Republic of Liberia

ACI Agence Congolaise d'Information, Brazzaville, People's Republic of the Congo

ACP Agence Camerounaise de Presse, Yaounde, Republic of Cameroon

ACP-C Agence Centrale de Presse-Communication, Paris, French Republic

ADN Allgemeiner Deutscher Nachrichtendienst, Berlin, Federal Republic of Germany

ADP Agence Dahoméenne de Presse, Cotonou, Republic of Benin

AFP Agence France-Press, Paris, French Republic

Agence Belga Agence Télégraphique Belge de Presse, Brussells, Kingdom of Belgium

Agence Centrafricaine de Presse Bangui, Central African Republic

Agence Telegraphic Monte Carlo, Principality of Monaco

Agencia Informativa Centroamericana Guatemala City, Republic of Guatemala

Agencia INNAC Santiago, Republic of Chile

Agencia Intercontinental de Impresa São Paulo, Federative Republic of Brazil

Agencia Meridional Rio de Janeiro, Federative Republic of Brazil

Agencia Orbe Latinoamericana Santiago, Republic of Chile

Agencja Publicystyczno   Warsaw, Republic of Poland

Agerpress   Agentia Romana de Presa, Bucharest, Republic of Romania

AGP   Agence Guinéenne de Presse, Conakry, Republic of Guinea

AIG   Agence d'Information Gabonaise, Libreville, Gabonese Republic

AIP   Agence Ivorienne de Presse, Abidjan, Côte d'Ivoire

AKP   Agence Kmere de Presse, Phnom Penh, Republic of Cambodia

AN   Agencia Nacional, Rio de Janeiro, Federative Republic of Brazil

Anatolia   Anatolia Ajansi, Ankara, Republic of Turkey

ANI   Agence National d'Information, Beirut, Republic of Lebanon

ANI   Agencia Nacional de Informaciónes, Montevideo, Oriental Republic of Uruguay

ANIM   Agence National d'Information Malienne, Bamako, Republic of Mali

ANOP   Agencia Noticiosa Portuguesa, Lisbon, Republic of Portugal

ANP   Algemeen Nederlands Presbureau/Netherlands Press Agency, Den Haag, Kingdom of the Netherlands

ANS   Agencia Noticiosa Saporiti, Buenos Aires, Argentine Republic

ANSA   Agenzia Nazionale Stampa Associata, Rome, Italian Republic

Antara   Indonesian National News Agency, Djakarta, Republic of Indonesia

AP   Associated Press, The, New York, N.Y., United States of America

APA   Austria Presse Agentur, Vienna, Republic of Austria

APC   Agence Presse Congolaise, Kinshasa, Republic of Zaire

APN   Agentsvo Pocati Novosti, Moscow, Russian Federation

APP   Associated Press of Pakistan, Karachi, Islamic Republic of Pakistan

APS   Algerie Presse Service, Algiers, Democratic and Popular Republic of Algeria

APS   Agence de Presse Senegalaise, Dakar, Republic of Senegal

APTV   Associated Press Television (International), London, United Kingdom of Great Britain and Northern Ireland

APV   Agence de Presse Voltaique, Ouagadougou, Burkina Faso

AR   Agencja Robatnicza, Warsaw, Republic of Poland

Argus Press   Rio de Janeiro, Federative Republic of Brazil

ATA   Agence Télégraphic Albanaise, Tirana, People's Socialist Republic of Albania

Atlantic News   Rio de Janeiro, Federative Republic of Brazil

ATP   Agence Tobadienne de Presse, Fort-Lamy, Republic of Chad

ATS   Agence Télégraphique Suisse, Bern, Swiss Federation. See also *SDA*.

AUP   Australian United Press, Sydney, Commonwealth of Australia

Bakhtar   Bakhtar New Agency, Kabul, Republic of Afghanistan

Bernama   Bernama News Agency, Kuala Lumpur, Malaysia

BIP   Brussels International Press Agency, Brussels, Kingdom of Belgium

BNS   Baltic News Service, Vilnius, Republic of Lithuania

BTA   Bulgarska Telegrafna Agencia, Sofia, Republic of Bulgaria

CAF   Centralna Agencja Fotograficzna, Warsaw, Republic of Poland

Calans   Caribbean and Latin American News Service, San Juan, Puerto Rico

CanP   Canadian Press, Toronto, Ontario, Canada

China News   China News and Publication Service, Taipei, Republic of China

China Union Press   Taipei, Republic of China

CIP   Centre d'Information de Presse, Agence d'Information Religieuse, Brussels, Kingdom of Belgium

CNA   Central News Agency, Taipei, Republic of China

Colombia Press   Bogota, Republic of Colombia

COPER   Agencia Noticiosa Corporacion de Periodistas, Santiago, Republic of Chile

CTK   Ceska Tiskova Kancelar, Prague, Czech Republic

Daily News Agency   Taipei, Republic of China

DDP   Deutscher Depeschen-Dienst, Bonn, Federal Republic of Germany

Dhiman Press of India   Ludhiana, Punjab, Republic of India

DLP   Demokraattinen Lehdistopalvelu, Helsinki, Republic of Finland

Donghwa   Donghwa News Agency, The, Seoul, Republic of Korea

DPA   Deutsche Presse Agentur, Hamburg, Federal Republic of Germany

EFE   Agencia EFE, Madrid, Kingdom of Spain

E.I.S.   Europe Information Services, Brussels, Kingdom of Belgium

ELTA   Lithuanian News Agency, Vilnius, Republic of Lithuania

Empresa Jornalistica Transpress   Rio de Janeiro, Federative Republic of Brazil

ENA   Ethiopian News Agency, Addis Ababa, People's Democratic Republic of Ethiopia

ETA   Estonian Telegraphic Agency, Tallin, Republic of Estonia

EXTEL   Exchange Telegraph Co., London, United Kingdom of Great Britain and Northern Ireland

Federal News Service   Washington, D.C., United States of America

FENA   Far East News Agency, The, Taipei, Republic of China

Fides Service   Agenzia Internazionale Fides, Vatican City State

GNA   Ghana News Service, Accra, Republic of Ghana

HA   Haber Ajansi, Istanbul, Republic of Turkey

Hapdong   Hapdong News Agency, Seoul, Republic of Korea

Hrvatska Izvjestajna Novinska Agencija/Croatian News & Press Agency   Zagreb, Republic of Croatia

HS   Hindustan Samachar, New Delhi, Republic of India

IANA   Inter-African News Agecy, Salisbury, Zimbabwe

IINA   Ireland International News Agency, Dublin, Ireland

INA   Iraqi National Agency, Baghdad, Republic of Iraq

INA   Israel News Agency, Tel Aviv, State of Israel

Informex   Mexico, DF, United Mexican States

Interpress   Bogota, Republic of Colombia

IRNA   Islamic Republic News Agency, Teheran, Islamic Republic of Iran

ITAR-TASS   International Telegraphic Agency of Russia. See *TASS*.

ITIM   News Agency of the Associated Israel Press, Tel Aviv, State of Israel

ITN   Independent Television News Service, London, United Kingdom of Great Britain and Northern Ireland

Japan Press, The   Tokyo, Japan

Jordanian News Agency   Amman, Hashemite Kingdom of Jordan

JP   Jiji Press, The, Tokyo, Japan

JTA   Jewish Telegraphic Agency, Jerusalem, State of Israel

KHS   Kaigai Hyoron Sho, Tokyo, Japan

KIPA   Katolische Internationale Presse Agentur, Zurich, Swiss Federation

KNA   Kenya News Agency, Nairobi, Republic of Kenya

Korea Central News Agency   Pyongyang, Democratic People's Republic of Korea

KPL   Khaosan Pathet Lao, Vientiane, Lao People's Democratic Republic

Kyoto   Kyoto News Service, The, Tokyo, Japan

Liberian Information Service   Monrovia, Republic of Liberia

LNA   Libyan News Agency, Tripoli, Socialist People's Libyan Arab Jamahiriya

Lusitania   Agencia Noticiosa de Luis C. Lupi, Lisbon, Republic of Portugal

Mad Presse   Madagascar Presse, Tananarive, Democratic Republic of Madagascar

MAP   Maghreb Arabe Presse, Rabat, Kingdom of Morocco

MBNA   Min Ben News Agency, Taipei, Republic of China

MENA   Middle East News Agency, Cairo, Arab Republic of Egypt

MNA   Myanmar News Agency, Yangon, Union of Myanmar

Monzame   Mongolian Press Agency, Ulan-Bator, Mongolian People's Republic

MTI   Magyar Távirati Iroda, Budapest, Republic of Hungary

Naewoe Press   Seoul, Republic of Korea

National News Service   Manila, Republic of the Philippines

Noti-Nica   Noticias Nicaragua, Managua, Republic of Nicaragua

NP   Noticias de Portugal, Lisbon, Republic of Portugal

NPA Namibian Press Agency, Windhoek, Republic of Namibia

NTB Norsk Telegrambyra, Oslo, Kingdom of Norway

NZPA New Zealand Press Association, Wellington, New Zealand

OP Orient Press, Seoul, Republic of Korea

PA Press Association, The, Ltd., London, United Kingdom of Great Britain and Northern Ireland

Pananews Pan-Asia Newspaper Alliance, Hong Kong

PAP Polska Agencja Prasowa, Warsaw, Republic of Poland

Pars Agency Pars News Agency, Teheran, Islamic Republic of Iran

PAT Press Association of Thailand, Bangkok, Kingdom of Thailand

PEVE Prensa Venezolana, Caracas, Republic of Venezuela

Philippine Press International Manila, Republic of the Philippines

PNS Philippine News Service, Manila, Republic of the Philippines

PPI Pakistan Press International, Karachi, Islamic Republic of Pakistan

PRELA Prensa Latina, Havana, Republic of Cuba

Press Trust of Ceylon Colombo, Democratic Socialist Republic of Sri Lanka

PRYC Agencia Noticiosa Prensa Radio y Cine, Santiago, Republic of Chile

PS Press Servis, Belgrade, Federal Republic of Yugoslavia

PTI Press Trust of India, Bombay, Republic of India

RAIS Rossiskoje Agentstvo Intellektual'noj Sobstvennosti Pri Prisidente Rossiskoj Federacii, Moscow, Russia.

Raos Press Features Agency Bangalore, Republic of India

RB Ritzaus Bureau, Copenhagen, Kingdom of Denmark

RITA Russian Information Telegraph Agency, Moscow, Russia

RN Reuters News Agency, London, United Kingdom of Great Britain and Northern Ireland

RSS Rashtriya Sambad Samiti, Katmandu, Kingdom of Nepal

Rudipresse Bujumbura, Republic of Burundi

SANA Syrian Arab News Agency, Damascus, Syrian Arab Republic

SAPA South African Press Association, Johannesburg, Republic of South Africa

Schweizerische Politische Korrespondenz Bern, Swiss Confederation

SDA Schweizerische Depeschenagentur, Bern, Swiss Confederation. See also *ATS.*

Servicio Informativo Continental y Orbe Latinoamericana Buenos Aires, Argentine Republic

Servizio Stampa Vatican City State

SICO Servizio Informazione Chiesa Orientale—Vatican City State

SIP Swenska Internationella Pressbyran, Stockholm, Kingdom of Sweden

SONNA Somali National News Agency, Mogadishu, Somali Democratic Republic

South Pacific News Service Wellington, New Zealand

SP Servicio Nacional de Prensa, Bogota, Republic of Colombia

STA Slovenska Tiskovna Agencija, Ljubljana, Republic of Slovenia

Star Star (Muslim) News Agency, Karachi, Islamic Republic of Pakistan

STT-FNB Oy Suomen Tietotoimisto-Finska Notisbyran, Helsinki, Republic of Finland

Sudanese Press Agency Khartoum, Republic of the Sudan

SUNA Sudan News Agency, Khartoum, Republic of the Sudan

Tanjug Telegrafska Agencija Nova Jugoslavia, Belgrade, Federal Republic of Yugoslavia

TAP Tunis Afrique Presse, Tunis, Republic of Tunisia

TA-SR Tlakova Agentura Slovaskej Republiky, Bratislava, Slovak Republic

TASS Telegrafnoje Agentstvo Suverenykh Stran/Telegraphic Agency of the Sovereign States, Moscow, Russia

TELAM Telenoticiosa Americana, Buenos Aires, Argentine Republic

Telpress Buenos Aires, Argentine Republic

THA Turk Haberler Ajansi, Istanbul, Republic of Turkey

Tonga Broadcasting News Service   Nuku'alofa, Kingdom of Tonga

TP   Teleprensa, Bogota, Republic of Colombia

TT   Tidningarnas Telegrambyra, Stockholm, Kingdom of Sweden

TTNA   Ta Tao News Agency, Taipei, Republic of China

UK News   United Kingdom News, London, United Kingdom of Great Britain and Northern Ireland

UNI   United News of India, New Delhi, Republic of India

UP   Uutispalvelu, Helsinki, Republic of Finland

UPI   United Press International, Washington, D.C., United States of America

UPP   United Press of Pakistan, Lahore, Islamic Republic of Pakistan

Uutiskeskus/Maaseutulehtien Liitto r.y.   Helsinki, Republic of Finland

VENPRES   Caracas, Republic of Venezuela

VISNEWS   London, United Kingdom of Great Britain and Northern Ireland

VNA   Vietnam News Agency, Hanoi, Socialist Republic of Vietnam

Wireless News Agency   Manila, Republic of the Philippines

Wreh News Agency   Monrovia, Republic of Liberia

WTN   Worldwide Television News, London, United Kingdom of Great Britain and Northern Ireland

XINHUA   Xinhua News Agency, Beijing, People's Republic of China

Yonhap News Agency   Seoul, Republic of Korea

ZANA   Zambia News Agency, Lusaka, Republic of Zambia

ZAP   Zachodnia Agencja Prasowa, Poznan, Republic of Poland